A First Look At Perturbation Theory

by

James G. Simmonds
University of Virginia

and

James E. Mann, Jr.
Wheaton College

Robert E. Krieger Publishing Company
Malabar, Florida
1986

Original Edition 1986

Printed and Published by
ROBERT E. KRIEGER PUBLISHING COMPANY, INC.
KRIEGER DRIVE
MALABAR, FL 32950

Library of Congress Cataloging in Publication Data

Simmonds, James G.
  A first look at perturbation theory.

   1. Differential equations – Numerical solutions.
2. Approximation theory. 3. Perturbation (Mathematics)
I. Mann, James E. II. Title.
QA371.S46 1985   515.3′5   84-20181
ISBN 0-89874-816-X

10 9 8 7 6 5 4 3 2

To those we love

PREFACE

Perturbation theory is fun, useful, and, we believe, accessible to undergraduates in engineering and the physical sciences. The mathematical background we expect of you, the reader, is modest: a two-semester course in one-variable calculus and a one-semester course in ordinary differential equations. Our book is written in an informal style, stressing heuristics. We have tried always to move from specific examples to generalities, emphasizing the "why" along with the "how." Both of us have used the material in this book in classes, and we know that the ideas can be grasped if you work consciously with them. This means that you *should* read this book with pencil and paper at hand to perform some of the implied algebraic operations and that you *must* work most of the exercises. Perturbation theory, like any art, must be learned by doing. Fortunately, many of the tedious calculations in perturbation theory traditionally carried out by hand can now be relegated to the computer thanks to symbol manipulation programs, such as MACSYMA developed at the Laboratory for Computer Science at the Massachusetts Institute of Technology.

Since perturbation methods produce only approximate solutions, one may ask, "Why not use numerical methods?" One answer is that perturbation methods produce *analytic* approximations that often reveal the essential dependence of the exact solution on the parameters in a more satisfactory way than does a numerical solution. A second answer is that some problems which cannot be easily solved numerically may yield to perturbation methods. Indeed, numerical and perturbation methods should be combined in a complementary way.

Let us mention the organization of the book. Chapter I is an overture that introduces the major themes that are elaborated upon in the chapters that follow. Chapter II is a thorough, but not exhaustive, treatment of how to find the roots of polynomials whose coefficients contain a small parameter. This chapter introduces regular and singular perturbations in as simple a context as possible. Working through this chapter carefully will help you to fix the idea of using only algebraic manipulations to solve problems. Subsequent chapters concentrate on differential equations. Here, we introduce you to many techniques for handling perturbations which change the order of the equations or which work for differential equations whose independent variable is unbounded. We end the book with two disparate practical problems that can be solved efficiently with perturbation methods.

Many people have influenced this book, but it is with special warmth that we recall the lessons of two of our respective teachers – George

Carrier (jem) and Eric Reissner (jgs) – who introduced us to the power and delights of perturbation theory. And we remember the deep and dazzling mastery of the subject by our friend and former colleague, Gordon Latta.

Finally, our thanks to Carolyn Duprey whose efforts with the computer made multiple revisions of the manuscript possible.

CONTENTS

CHAPTER I:
INTRODUCTION AND OVERVIEW

Perturbation theory is the study of the effects of small disturbances. If the effects are small, the disturbances or *perturbations* are said to be *regular*; otherwise, they are said to be *singular*. The basic idea in perturbation theory is to obtain an approximate solution of a mathematical problem by exploiting the presence of a small dimensionless parameter – the smaller the parameter, the more accurate the approximate solution.

Regular perturbations are assumed nearly every time we construct a mathematical model of a real world phenomenon. Our choice of language reflects this: the flow is *almost* steady, the density varies *essentially* with altitude only, the conductivity is *virtually* independent of temperature, the spring is *nearly* linear, the friction is *practically* negligible.

Singular perturbations are probably less familiar. Fig. 1.1 illustrates two examples. The top in Fig. 1.1a is set spinning rapidly about a vertical axis. During one revolution, the effects of aerodynamic drag and friction at the tip are small (regular perturbations). Eventually, though, the top falls and comes to rest in a position far from its initial one. Thus, over a long period of time, the perturbations are singular. Models such as this are characterized by what may be called *a singularity in the domain*. For the top, this means that we are interested in what happens *for all time* after its release, that is, for all times t on the semi-infinite (and therefore singular) domain $t > 0$.

Fig. 1.1b shows a hemispherical elastic shell under an internal pressure p. The shell has been clamped at its edge, which prevents displacement or rotation there. Since the shell is symmetric, the stresses depend only on the polar angle ϕ. The maximum stress occurs in the outer fibers of the shell and can be found by solving non-homogeneous ordinary differential equations. The solutions contain constants determined by regularity conditions at the pole ($\phi = 0$) and edge conditions at the equator ($\phi = \pi/2$). The differential equations contain the small parameter $h//R$,

1

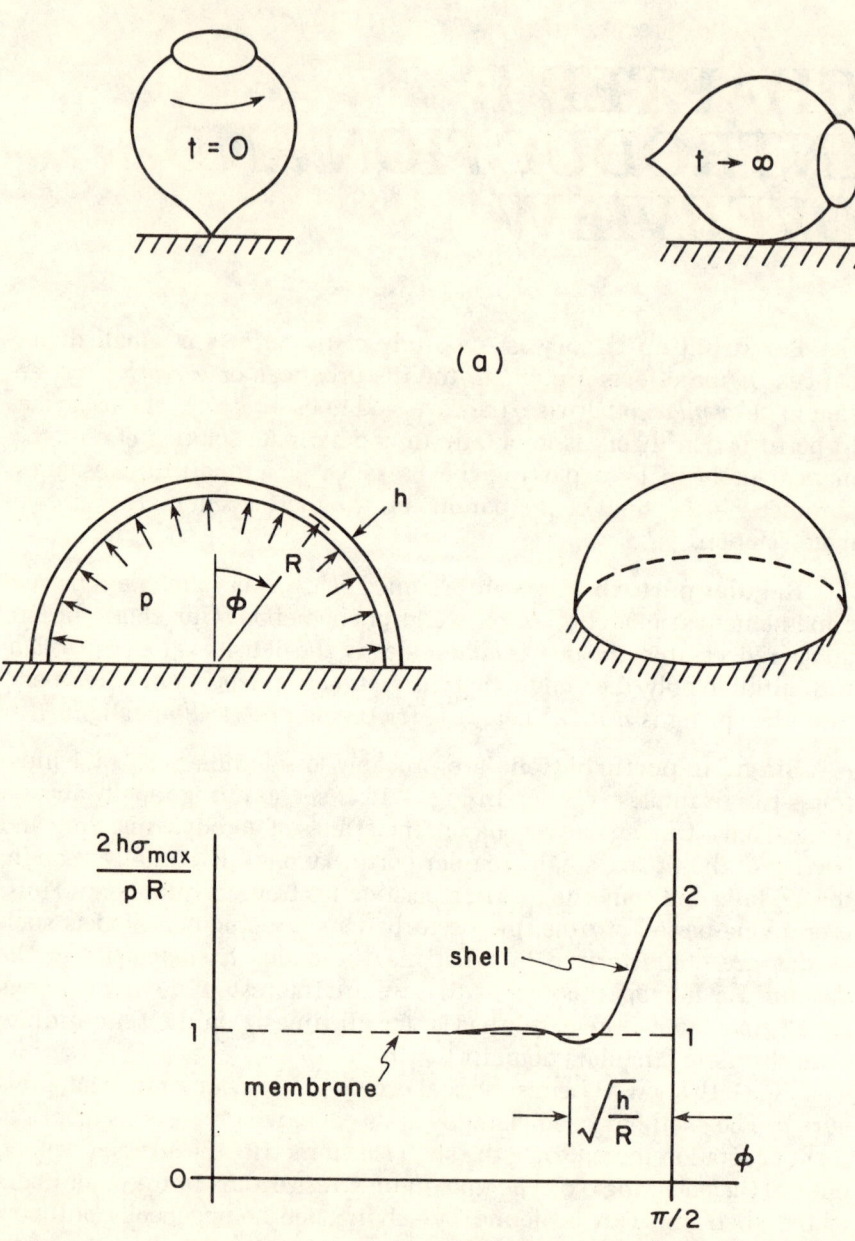

Fig. 1.1. Singular perturbations resulting from (a) a singularity in the domain and (b) a singularity in the model.

where h is the shell thickness and R is the radius of the shell midsurface. If h/R is set to zero in these equations, then we obtain the equation for a *membrane, i.e.,* a shell with no bending stiffness. The solutions of these simplified equations predict a maximum stress of $pR/2h$, everywhere.[1] The defect of these simplified equations is that their solutions cannot meet the edge conditions.

In Fig. 1.1b we have plotted, as a function of ϕ, the maximum dimensionless stress $(2h\sigma/pR)$ as predicted by shell and membrane theory for a typical value of h/R. Except near the equator, the results are virtually identical. However, in a narrow zone near the equator – the *boundary layer* – the dimensionless maximum stress predicted by shell theory dips below 1 and then rises to a value of 2. The key feature of this graph is that no matter how small the parameter h/R, the rise of the interior stress by a factor of 2 at the edge never diminishes. This is a singular perturbation phenomenon. The "shell versus membrane" solutions reflect what may be called a *singularity in the model.* Setting $h/R = 0$ leads to an over-simplified model that fails to predict the non-negligible stress rise at the boundary. This failure of the membrane theory occurs in a narrow region near the boundary; the width of the failure region depends on the size of h/R.

Perturbation problems arising from a singularity in the domain were first studied systematically by Poincaré, who encountered them in orbital mechanics. The first extensive analysis of problems involving a singularity in the model (boundary layer problems) was carried out by Prandtl in his work on the flow over solid objects of fluids of low viscosity such as air and water. Although the problems attacked by Poincaré and Prandtl are too elaborate for this book, we can explore many aspects of perturbation theory by working with simple equations, many of which can be applied to common phenomena.

The First Quantitative Step. We begin with the following problem from the theory of quadratic equations:

> Determine how the roots of $z^2 - 2z + \epsilon$ change as ϵ
> is perturbed slightly away from zero.

If $\epsilon = 0$, the roots are, by inspection, $z_1 = 0$ and $z_2 = 2$. For other values of ϵ we have, as a result of the quadratic formula,

$$z_1(\epsilon) = 1 - \sqrt{1 - \epsilon} \qquad (1.1)$$

---

[1]This result is also derived easily if we divide a complete spherical membrane by an imaginary equatorial plane and then equate the pressure force, $p\pi R^2$, to the tension along the equator, $2\pi Rh\sigma$, where σ is the stress.

$$z_2(\epsilon) = 1 + \sqrt{1 - \epsilon}. \tag{1.2}$$

The Numerics of a Regular Problem. Using a hand calculator[2], we readily construct from (1.1) and (1.2) the following table.

Table 1.1. Roots of $z^2 - 2z + \epsilon$

ϵ	$z_1(\epsilon)$	$z_2(\epsilon)$
10^{-1}	$0.0513167 \cdots$	$1.9486833 \cdots$
10^{-2}	$0.0050126 \cdots$	$1.9949874 \cdots$
$\vdots$	$\vdots$	$\vdots$
10^{-6}	$0.0000005 \cdots$	$1.9999995 \cdots$
10^{-7}	5.01×10^{-8}	$1.9999999 \cdots$

Our numerical calculations suggest that a perturbation about $\epsilon = 0$ is regular but teach us little else. Moreover, we took no advantage of the fact that the roots for $\epsilon = 0$ came with little effort. We shall remedy this lack of analysis presently.

The Numerics of a Singular Problem. If we switch ϵ with the coefficient of z^2, we have the polynomial $\epsilon z^2 - 2z + 1$. If $\epsilon = 0$, $z_1 = \frac{1}{2}$ is its only root. If $\epsilon \neq 0$, there are two:

$$z_1(\epsilon) = \frac{1 - \sqrt{1 - \epsilon}}{\epsilon} \tag{1.3}$$

$$z_1(\epsilon) = \frac{1 + \sqrt{1 - \epsilon}}{\epsilon}. \tag{1.4}$$

Again, using a hand calculator, we construct the following table.

Table 1.2. Roots of $\epsilon z^2 - 2z + 1$

ϵ	$z_1(\epsilon)$	$z_2(\epsilon)$
10^{-1}	$0.5131670 \cdots$	$19.486833 \cdots$
10^{-2}	$0.5012563 \cdots$	$199.49874 \cdots$
$\vdots$	$\vdots$	$\vdots$
10^{-6}	$0.5001000 \cdots$	$1999999.5 \cdots$
10^{-7}	$.501 \cdots$	19999999

What is notable about Table 1.2? First, while one of the roots, $z_1(\epsilon)$,

---

[2]A Texas Instruments TI-30-II in our case.

approaches that of $2z - 1$ as $\epsilon \to 0$, the other, $z_2(\epsilon)$, goes to infinity. This is a manifestation of singular behavior. On a more fundamental level, *changing ϵ from 0 to an arbitrarily small number has changed the number of solutions for z* (from one to two). Notice that the equation changes from linear ($\epsilon = 0$) to quadratic ($\epsilon \neq 0$). Such a change in the order of an equation characterizes many singular perturbation problems.

A second feature of Table 1.2 is that as $\epsilon \to 0$, $z_1(\epsilon)$ at first approaches .5 steadily but then, for very small values of ϵ, begins to exhibit small fluctuations. These are caused by round-off errors produced by computing differences of nearly equal terms in (1.3). While there is a simple way to remedy this in the present problem – multiply numerator and denominator in (1.3) by $1 + \sqrt{1 - \epsilon}$ – the diagnosis of round-off error in more complicated problems, much less the cure, is not so simple. In cases such as these in which numerics falter, perturbation theory can sometimes save the day.

Analysis of the Regular Problem. To find approximate solutions that are accurate and easy to use, we study the effect of a small parameter. First, let us find approximate formulas for the roots of $z^2 - 2z + \epsilon$ when ϵ is small. Unlike most perturbation problems, this one can be analyzed completely. We analyze precisely the simplest problem of a class with the hope of inferring patterns or principles which can aid in the attack on more complicated problems.

Infinite Series. The formulas (1.1) and (1.2) for the roots of $z^2 - 2z + \epsilon$ are exact. However, the smallness of ϵ simplifies the problem. We can certainly assume that $|\epsilon| < 1$. This not only rules out complex roots, but, more importantly, it allows us to expand $\sqrt{1 - \epsilon}$ in *a power series in ϵ*. Recall the binomial expansion:

$$(a+b)^m = a^m + ma^{m-1}b + \frac{m(m-1)}{2} a^{m-2}b^2 + \cdots + \frac{m!}{k!(m-k)!} a^{m-k}b^k + \cdots . \quad (1.5)$$

Setting $a = 1$, $b = -\epsilon$, and $m = \frac{1}{2}$, we have

$$\sqrt{1 - \epsilon} = 1 - \frac{1}{2}\epsilon - \frac{1}{8}\epsilon^2 - \cdots . \quad (1.6)$$

This is a *formal* series, so-called because we as yet have made no attempt to ask what it means to try to add together an infinite number of terms.

The way to study an infinite series is to study its sequence of partial sums. If the sequence coverges, then we say that the series converges, and if we are able to compute the limit, S, of this sequence, we say that the *series sums to S*. One of the simplest tests for convergence is the ratio test, which states that Σu_k converges if

$$\lim_{k \to \infty} \left| \frac{u_{k+1}}{u_k} \right| < 1.$$

From (1.5) and (1.6) we find that

$$\lim_{k \to \infty} \left| \frac{(k+1)^{st} \text{ term}}{k^{th} \text{ term}} \right| = \lim_{k \to \infty} \left| \frac{k - \frac{1}{2}}{k+1} \epsilon \right| = | \epsilon |. \tag{1.7}$$

Thus the right side of (1.6) converges if $| \epsilon | < 1$, as do the two series

$$z_1(\epsilon) = \frac{1}{2}\epsilon + \frac{1}{8}\epsilon^2 + \cdots$$

$$z_2(\epsilon) = 2 - \frac{1}{2}\epsilon - \cdots \tag{1.8}$$

that we obtain upon substituting (1.6) into (1.1) and (1.2).

The first few terms of these series must, to be useful, yield close approximations to the roots. Taking only those terms displayed explicitly in (1.8) and $\epsilon = 0.1$, we find $z_1(.1) \approx .5(.1) + .125(.1)^2 = .05125$ and $z_2(.1) \approx 2 - .5(.1) = 1.95$. The exact values are $z_1(.1) = .05131 \ldots$ and $z_2(.1) = 1.94 \ldots$. Are these roots accurate enough? You must decide.

In a more elaborate problem, there may not be a formula for the exact solution(s) with which one can check the accuracy of an approximation, or, if there is an exact formula, it may be difficult to apply. As an example of the latter, consider finding the roots of

$$z^3 - 5z^2 + 4z + \epsilon = 0 \tag{1.9}$$

when ϵ is small. Though there is an exact formula for the roots of this (and any other) cubic, it is difficult to use. However, if $\epsilon = 0$, we have

$$z^3 - 5z^2 + 4z = 0 \tag{1.10}$$

which has roots

$$z_1(0) = 0$$

$$z_2(0) = 1$$

$$z_3(0) = 4. \tag{1.11}$$

Presently, we shall develop a technique which exploits the smallness of ϵ to produce an acceptable approximation of the roots of any polynomial $P_\epsilon(z)$ when the roots of the "nearby" or *reduced* polynomial $P_0(z)$ are known. But first we need a way of making error estimates that depends on the approximation process itself.

Taylor's expansion with a remainder uses the value of a function and its first few derivatives at a point at which they are easily found to estimate the value of the function at a nearby point where the function is difficult to calculate. More precisely, if a function $f(\epsilon)$ and its first n derivatives are continuous on a closed interval $|\epsilon| \leq a$, and if the $(n+1)$st derivative of $f(\epsilon)$ exists on the open interval $|\epsilon| < a$, then

$$f(\epsilon) = f(0) + f'(0)\epsilon + \cdots + \frac{f^{(n)}(0)}{n!} \epsilon^n + R_{n+1}(\epsilon). \quad (1.12)$$

Here $f^{(n)}(0)$ denotes the nth derivative of $f(\epsilon)$ evaluated at $\epsilon = 0$ and *the remainder after $n+1$ terms*, $R_{n+1}(\epsilon)$, has the form

$$R_{n+1}(\epsilon) = \frac{f^{(n+1)}(x)}{(n+1)!} \epsilon^{n+1}, \; |x| < |\epsilon|. \quad (1.13)$$

The number x depends on ϵ but is otherwise unknown.

The important facts about the expansion (1.12) are:

(1) it allows us to approximate $f(\epsilon)$ by a (Taylor) polynomial in ϵ, namely the right side of (1.12) without $R_{n+1}(\epsilon)$.

(2) if we can find an upper bound on $R_{n+1}(\epsilon)$ then we have an upper bound on the error of our polynomial approximation.

(3) it is almost incidental in applications whether the associated infinite series $f(0) + f'(0)\epsilon + \frac{1}{2} f''(0)\epsilon^2 + \cdots$ converges. The size of $R_{n+1}(\epsilon)$ is the salient fact.

With $f(\epsilon) = (1 - \epsilon)^{1/2}$, we have, if $n = 2$,

$$(1 - \epsilon)^{1/2} = 1 - \frac{1}{2}\epsilon - \frac{(1-x)^{-3/2}}{8} \epsilon^2, \; |x| < |\epsilon| < 1 \quad (1.14)$$

and, if $n = 3$,

$$(1 - \epsilon)^{1/2} = 1 - \frac{1}{2}\epsilon - \frac{1}{8}\epsilon^2 - \frac{(1-x)^{-5/2}}{16} \epsilon^3, \; |x| < |\epsilon| < 1. \quad (1.15)$$

Substituting (1.15) into (1.1) and (1.14) into (1.2), we find that

$$z_1(\epsilon) = \frac{1}{2}\epsilon + \frac{1}{8}\epsilon^2 + \frac{(1-x)^{-5/2}}{16} \epsilon^3, \; |x| < |\epsilon| < 1$$

$$(1.16)$$

$$z_2(\epsilon) = 2 - \frac{1}{2}\epsilon - \frac{(1-x)^{-3/2}}{8} \epsilon^2, \; |x| < |\epsilon| < 1.$$

These expressions offer a way of estimating the errors we make when truncating the infinite series expansions for $z_1(\epsilon)$ and $z_2(\epsilon)$ after two terms.

If $p > 0$, and $|x| < 1$, then $(1 - x)^{-p}$ is largest when x is as close to 1 as possible. Thus, if $|x| < 1 - \delta$, where δ is any fixed number such that $0 < \delta < 1$,

$$\left| \frac{(1 - x)^{-5/2}}{16} \epsilon^3 \right| < \frac{1}{16\delta^{5/2}} |\epsilon|^3.$$

$$\left| \frac{(1 - x)^{-3/2}}{8} \epsilon^2 \right| < \frac{1}{8\delta^{3/2}} |\epsilon|^2. \qquad (1.17)$$

Exercise 1.1. Earlier, we used the terms displayed explicitly in (1.8) to approximate $z_1(.1)$ and $z_2(.1)$. Use (1.16) and (1.17) to obtain an upper bound on the errors we made.

The Order Symbols. Using (1.17), we may rewrite (1.16) in the form

$$z_1(\epsilon) = \frac{1}{2}\epsilon + \frac{1}{8}\epsilon^2 + O(\epsilon^3) \qquad (1.18)$$

$$z_2(\epsilon) = 2 - \frac{1}{2}\epsilon + O(\epsilon^2). \qquad (1.19)$$

The symbols $O(\epsilon^3)$ and $O(\epsilon^2)$, to be read "big 'O' of ϵ cubed" and "big 'O' of ϵ squared" are used to sweep all irrelevant algebraic details under the rug.

In general, $g(\epsilon) = O(\epsilon^p)$ means that, for ϵ sufficiently small, there exists a positive constant K, *independent* of ϵ, such that $|g(\epsilon)| < K|\epsilon|^p$. In (1.18) and (1.19), "sufficiently small" means $|\epsilon| < 1 - \delta$, and the K's are $(1/16)\delta^{-5/2}$ and $(1/8)\delta^{-3/2}$ respectively. In more complicated problems, however, it is rare when we can pin down the words "sufficiently small" and "there exists." Thus, in practice, we may have to view a statement such as $f(\epsilon) = O(\epsilon^2)$ as simply implying that f grows no faster than the square of ϵ when ϵ is small.

Analysis of the Singular Problem. With what we have learned, we can now quickly analyze the singular problem of finding simple, approximate formulas for the roots of $\epsilon z^2 - 2z + 1$ when ϵ is small. Substituting the Taylor formula for $\sqrt{1 - \epsilon}$ into (1.3) and (1.4), we obtain

$$z_1(\epsilon) = \frac{1}{2} + \frac{1}{8}\epsilon + O(\epsilon^2) \qquad (1.20)$$

$$z_2(\epsilon) = \frac{2}{\epsilon} - \frac{1}{2} + O(\epsilon). \qquad (1.21)$$

The terms displayed give an approximation to $z_1(10^{-6})$ of $.5 + .125(10^{-6})$ $= .5000000125$, and an approximation to $z_2(10^{-6})$ of $2(10^6) - .5 = 1999999.5$. The approximation to $z_1(10^{-6})$ has been improved dramatically from the value found earlier with a calculator, but $z_2(\epsilon)$ still approaches infinity as ϵ approaches zero. *This behavior is inherent in the problem and is not a numerical artifact.* Our analysis of $z_2(\epsilon)$ has, nevertheless, provided useful information: we now know that $z_2(\epsilon)$ behaves like $2/\epsilon$ as $\epsilon \to 0$. For larger values of ϵ we might need more terms in the Taylor polynomial for $z_1(\epsilon)$ and $z_2(\epsilon)$ to obtain sufficiently accurate approximations.

Exercise 1.2 Make upper bound estimates of the remainders and determine the smallest Taylor polynomials that can produce approximations to the roots of $\epsilon z^2 - 2z + 1$ with an absolute error $< 10^{-3}$ for all $|\epsilon| < .2$.

Note that when we finally obtained useful numerical formulas for the roots of $z^2 - 2z + \epsilon$—(1.18) and (1.19)—each was of the form

$$z(\epsilon) = a_0 + a_1\epsilon + a_2\epsilon^2 + \cdots + a_N\epsilon^N + R_{N+1}(\epsilon) , \quad R_{N+1}(\epsilon) = O(\epsilon^{N+1}).$$
$$(1.22)$$

The right side of (1.22) is called an *asymptotic* or *regular perturbation expansion*. It is ideal for assessing, numerically or theoretically, the effect of a small perturbation in ϵ about zero. Though an asymptotic expansion need not converge as $N \to \infty$, we do require that $\epsilon^{-N}R_N \to 0$ as $\epsilon \to 0$ for fixed N. Any function with a representation of the form (1.22) is called *regular* because $z(\epsilon)$ approaches a finite value as ϵ approaches 0.

Reanalysis of the Regular Problem. Suppose we pretend that the quadratic formula does not exist. Let us assume instead that each root of $P_\epsilon(z) \equiv z^2 - 2z + \epsilon$ has a representation of the form (1.22) for some fixed integer N. Then we may attempt to determine the unknown coefficients $a_0, a_1 \cdots$ by requiring that

$$P_\epsilon(z(\epsilon)) = P_\epsilon(a_0 + a_1\epsilon + \cdots + a_N\epsilon^N + O(\epsilon^{N+1}))$$
$$= (a_0 + a_1\epsilon + \cdots + a_N\epsilon^N + O(\epsilon^{N+1}))^2 \qquad (1.23)$$
$$- 2(a_0 + a_1\epsilon + \cdots) + \epsilon = 0,$$

identically in ϵ as $\epsilon \to 0$. In this problem we know that N may be *any* positive integer, that $a_0 + a_1\epsilon + a_2\epsilon^2 + \cdots$ is a convergent power series, and that its radius of convergence is 1. However, as we shall see, such information is *not* necessary to determine the unknown coefficients in (1.22). Indeed, there are problems involving a small parameter in which the solutions are of the form (1.22), but in which the associated infinite series $a_0 + a_1\epsilon + \cdots$ does *not* converge for *any* non-zero value of ϵ. [For example, see Eq. (1.74).] Therefore, it is essential to emphasize that *the subsequent procedure does not involve infinite series.*

Before substituting the right side of (1.22) into (1.23), we must square it. Thus, if $z(\epsilon) = a_0 + R_1(\epsilon)$, where $R_1(\epsilon) = O(\epsilon)$, then $z^2(\epsilon) = a_0^2 + 2a_0R_1(\epsilon) + R_1^2(\epsilon)$. Now $2a_0R_1(\epsilon) = O(\epsilon)$, because $R_1(\epsilon) = O(\epsilon)$ implies that there exists a constant K such that $|R_1(\epsilon)| < K|\epsilon|$ for ϵ sufficiently small. Hence, $|2a_0R_1(\epsilon)| < 2|a_0K||\epsilon|$ is true for ϵ sufficiently small. Furthermore, $R_1^2(\epsilon) = O(\epsilon^2)$, because $|R_1(\epsilon)| < K|\epsilon|$ implies that $|R_1^2(\epsilon)| < K^2|\epsilon|^2$. In fact, since $K^2|\epsilon|^2 < K^2|\epsilon|$ if $|\epsilon| < 1$, we can also write $R_1^2(\epsilon) = O(\epsilon)$. Finally, it is important to note that the sum of two $O(\epsilon)$-terms is again $O(\epsilon)$. For if there exist constants P and Q such that $|p(\epsilon)| < P|\epsilon|$ and $|q(\epsilon)| < Q|\epsilon|$ for ϵ sufficiently small, then $|p(\epsilon) + q(\epsilon)| < K|\epsilon|$, where $K = P + Q$. In summary, we may conclude that if $z(\epsilon) = a_0 + O(\epsilon)$, then $z^2(\epsilon) = a_0^2 + O(\epsilon)$. See what a marvelous invention the O-symbols are!

Again, if $z(\epsilon) = a_0 + a_1\epsilon + R_2(\epsilon)$, where $R_2(\epsilon) = O(\epsilon^2)$, then

$$z^2(\epsilon) = a_0^2 + 2a_0a_1\epsilon + a_1^2\epsilon^2 + 2a_0R_2(\epsilon) + 2a_1\epsilon R_2(\epsilon) + R_2^2(\epsilon)$$

(1.24)

$$= a_0^2 + 2a_0a_1\epsilon + O(\epsilon^2).$$

To understand how we could make such a simplifying leap, let us take, one by one, the terms in the first line of (1.24) that we swept into the O-symbol in the second line. If K is any constant greater than $|a_1^2|$, then $|a_1^2\epsilon^2| < K|\epsilon|^2$, i.e., $a_1^2\epsilon^2 = O(\epsilon^2)$. (To allow for the possibility that the unknown coefficients might be complex numbers, we write $|a_1^2|$ instead of a_1^2). The term $2a_0R_2(\epsilon)$ is also $O(\epsilon^2)$ because $|R_2(\epsilon)| < K|\epsilon|^2$ implies that $|2a_0R_2(\epsilon)| < 2|a_0|K|\epsilon|^2$. For the same reason, $2a_1\epsilon R_2(\epsilon) = O(\epsilon^3)$. But a term that is $O(\epsilon^3)$ is also $O(\epsilon^2)$ because $K|\epsilon|^3 < K|\epsilon|^2$ if $|\epsilon| < 1$. Finally, as $|R_2(\epsilon)| < K|\epsilon|^2$ for ϵ sufficiently small, it follows that $|R_2^2(\epsilon)| < K^2|\epsilon|^4 < K^2|\epsilon|^2$ for ϵ sufficiently small. That is, $R_2^2(\epsilon)$ is both $O(\epsilon^4)$ *and* $O(\epsilon^2)$. Thus the last four terms in the first row of (1.24) are each $O(\epsilon^2)$. But, by the same argument we used earlier to show that the sum of two $O(\epsilon)$ terms is again $O(\epsilon)$, we conclude that a finite sum of $O(\epsilon^2)$ term is again $O(\epsilon^2)$, and thus arrive at the second line of (1.24).

Before proceeding further into perturbation theory, be sure that you have a clear understanding of how O-symbols work; they are essential to the rest of this book.

After computing explicitly one more term in the representation for $z^2(\epsilon)$, we can write

$$z^2(\epsilon) = a_0^2 + 2a_0a_1\epsilon + (a_1^2 + 2a_0a_2)\epsilon^2 + O(\epsilon^3). \qquad (1.25)$$

Substituting this expression and (1.22), with $N = 2$, into $P_\epsilon(z) = z^2 -$

$2z + \epsilon$, we obtain

$$P_\epsilon(z(\epsilon)) = a_0^2 + 2a_0a_1\epsilon + (a_1^2 + 2a_0a_2)\epsilon^2 + O(\epsilon^3)$$

$$- 2[a_0 + a_1\epsilon + a_2\epsilon^2 + O(\epsilon^3)] + \epsilon = 0. \tag{1.26}$$

Collecting coefficients of like powers of ϵ and noting that the sum of two terms of $O(\epsilon^3)$ is again a term of $O(\epsilon^3)$, we can rewrite (1.26) in the form

$$P_\epsilon(z(\epsilon)) = a_0^2 - 2a_0 + (2a_0a_1 - 2a_1 + 1)\epsilon$$

$$+ (a_1^2 + 2a_0a_2 - 2a_2)\epsilon^2 + O(\epsilon^3) = 0. \tag{1.27}$$

It is important to keep in mind that to say that (1.22) is a root of $P_\epsilon(z)$, for ϵ sufficiently small, means that $P_\epsilon(z(\epsilon))$ *must be identically zero* ($\equiv$ 0) *for all values of* ϵ sufficiently *small*. In particular, then, (1.27) must hold as $\epsilon \to 0$. Therefore, because the left side is a continuous function of ϵ,

$$a_0^2 - 2a_0 = 0. \tag{1.28}$$

This result, in turn, reduces (1.27) to

$$P_\epsilon(z(\epsilon)) = (2a_0a_1 - 2a_1 + 1)\epsilon + (a_1^2 + 2a_0a_2 - 2a_2)\epsilon^2 + O(\epsilon^3) = 0. \tag{1.29}$$

But if $P_\epsilon(z(\epsilon)) \equiv 0$ for ϵ sufficiently small, then $\epsilon^{-1}P_\epsilon(z(\epsilon)) \equiv 0$ for ϵ sufficiently small and *nonzero*. This implies that[3] $\lim \epsilon^{-1}P_\epsilon(z(\epsilon)) = 0$ which, by (1.29), yields

$$2a_0a_1 - 2a_1 + 1 = 0. \tag{1.30}$$

Thus (1.29) reduces still further to

$$P_\epsilon(z(\epsilon)) = (a_1^2 + 2a_0a_2 - 2a_2)\epsilon^2 + O(\epsilon^3) = 0. \tag{1.31}$$

Again, since (1.31) must hold for all values of ϵ sufficiently small, we conclude that $\lim \epsilon^{-2}P_\epsilon(z(\epsilon)) = 0$. This requires that

$$a_1^2 + 2a_0a_2 - 2a_2 = 0. \tag{1.32}$$

What we have just established, in a specific case, is

---

[3]"lim" is shorthand for $\lim_{\epsilon \to 0}$.

> ### The Fundamental Theorem of Perturbation Theory:
>
> If
> $$A_0 + A_1\epsilon + \cdots + A_N\epsilon^N + O(\epsilon^{N+1}) \equiv 0 \qquad (1.33)$$
>
> for ϵ sufficiently small and if the coefficients A_0, A_1, $\cdots$ are independent of ϵ, then
> $$A_0 = A_1 = \cdots = A_N = 0. \qquad (1.34)$$

To solve $z^2 - 2z + \epsilon = 0$, note that (1.28) has two roots that may be found by inspection:

$$a_0 = 0 \text{ and } a_0 = 2. \qquad (1.35)$$

Choose the first and (1.30) yields

$$a_1 = 1/2. \qquad (1.36)$$

Thus, from (1.32),

$$(1/2)^2 - 2a_2 = 0 \text{ or } a_2 = 1/8. \qquad (1.37)$$

Substituting these results into (1.22), we reproduce (1.18)—the representation for $z_1(\epsilon)$ obtained earlier from the quadratic formula.

If we make the second choice which satisfies (1.28), $a_0 = 2$, and follow the same steps, we find that $a_1 = -1/2$ and $a_2 = -1/8$. These results, substituted into (1.22), reproduce (1.19), the earlier representation for $z_2(\epsilon)$. The same procedures can be followed to produce as many terms of the series for $z_1(\epsilon)$ and $z_2(\epsilon)$ as desired.

Exercise 1.3 Use the above procedure to compute $z_1(\epsilon)$ and $z_2(\epsilon)$ when $N = 4$ in (1.22).

Merely knowing that a solution to a problem exists is usually of little help in finding it. However, an important principle is illustrated by our example. If we can learn something about the general *form* of a solution, say that it has a representation like (1.22), then the unknown parts of the solution may often be determined one at a time by direct substitution into the given equations.

Exercise 1.4. The roots of the polynomial $z^3 - z + \epsilon$ can be shown to be of the form (1.22). Find a_0 and a_1 for each root.

Re-analysis of the Singular Problem. Which roots of $Q_\epsilon(z) \equiv \epsilon z^2 - 2z + 1$ have the form (1.22) with $N = 1$? Substituting (1.22) along with (1.25) for $z^2(\epsilon)$ into $Q_\epsilon(z)$, we find that

$$Q_\epsilon(z(\epsilon)) = \epsilon[a_0^2 + 2a_0a_1\epsilon + O(\epsilon^2)] - 2[a_0 + a_1\epsilon + O(\epsilon^2)] + 1. \tag{1.38}$$

By collecting like powers of ϵ and requiring that $Q_\epsilon(z(\epsilon))$ be identically zero for ϵ sufficiently small, we obtain

$$-2a_0 + 1 + (a_0^2 - 2a_1)\epsilon + O(\epsilon^2) \equiv 0. \tag{1.39}$$

Using The *Fundamental Theorem of Perturbation Theory*, we set the coefficients of the various powers of ϵ in (1.39) to zero to obtain

$$-2a_0 + 1 = 0 \tag{1.40}_0$$

$$a_0^2 - 2a_1 = 0. \tag{1.40}_1$$

Solved in order, these equations yield

$$a_0 = 1/2 \tag{1.41}_0$$

$$a_1 = 1/8 \tag{1.41}_1$$

Substituting these values back into (1.22), we reproduce (1.20), the expression we obtained earlier using the quadratic formula. What about the other root, $z_2(\epsilon)$, given by (1.21)? Obviously, we did not pick it up because (1.21) is not of the form (1.22).

Recall that the point of our re-analysis is to pretend that we know neither the quadratic formula nor how to expand $\sqrt{1 - \epsilon}$ in an infinite series. With this in mind, we can still infer that *the missing root $z_2(\epsilon)$ must approach infinity as ϵ approaches zero*. Why? Because

$$Q_\epsilon(z) = \epsilon z^2 - 2z + 1$$
$$= \epsilon[z - z_1(\epsilon)][z - z_2(\epsilon)] \tag{1.42}$$
$$= \epsilon z^2 - \epsilon[z_1(\epsilon) + z_2(\epsilon)]z + \epsilon z_1(\epsilon)z_2(\epsilon).$$

Equating the 1 in the first line of (1.42) to $\epsilon z_1 z_2$ in the last line of (1.42) (since the two lines must be equal if $z = 0$), we see that

$$z_1(\epsilon)z_2(\epsilon) = \epsilon^{-1}. \tag{1.43}$$

This implies that as $\epsilon \to 0$, $z_2(\epsilon) \to \infty$, since $z_1(\epsilon)$, a regular root, must approach a finite value.

Now we can make a key simplification; when $z_2(\epsilon)$ is large, $-2z_2(\epsilon) + 1 \sim -2z_2(\epsilon)$. [When we write $f(\epsilon) \sim g(\epsilon)$, which is read "$f$ is asymptotic to g as ϵ approaches zero", we mean that $f/g \to$ a constant *as* $\epsilon \to 0$.] Thus

$$Q_\epsilon(z_2(\epsilon)) \sim \epsilon z_2^2(\epsilon) - 2z_2(\epsilon) \equiv H_\epsilon(z_2(\epsilon)). \tag{1.44}$$

By inspection, the roots of $H_\epsilon(z)$ are 0 and $2/\epsilon$. But since $z_2(\epsilon)$ gets large as ϵ gets small, it is evident that

$$z_2(\epsilon) \sim 2/\epsilon. \tag{1.45}$$

This result, obtained by *heuristics* (suggestive but not rigorous arguments), leads to the conjecture that

$$\epsilon z_2(\epsilon) \sim b_0 \neq 0. \tag{1.46}$$

At this point, the fact that $b_0 = 2$ is not essential; that $b_0 \neq 0$ is. This conjecture in turn suggests *the change of variable*

$$\epsilon z(\epsilon) = w(\epsilon). \tag{1.47}$$

Noting that $\epsilon Q_\epsilon(z) = \epsilon^2 z^2 - 2\epsilon z + \epsilon$, we have

$$\epsilon Q_\epsilon(z) = \epsilon Q(w/\epsilon) = w^2 - 2w + \epsilon \equiv R_\epsilon(w). \tag{1.48}$$

By (1.47) and (1.48), the roots of $R_\epsilon(w)$ give the roots of $Q_\epsilon(z)$. But as $R_0(z)$ has *two* roots, finding the roots of $R_\epsilon(w)$ for ϵ sufficiently small is a regular perturbation problem.

> The essence of singular perturbation theory is to extract (by heuristics, if necessary) the dominant singular behavior of a solution and then, by a change of variable, to reduce the singular problem to a regular one.

For the case at hand, we assume that each root of $R_\epsilon(w)$ has a regular perturbation expansion of the *form* (1.22) with, say, $N = 1$:

$$w(\epsilon) = b_0 + b_1\epsilon + O(\epsilon^2). \tag{1.49}$$

Substituting (1.49) into (1.48), collecting coefficients of like powers of ϵ and requiring that the resulting expression be identically zero for ϵ sufficiently small, we find

$$(b_0^2 - 2b_0) + (2b_0 b_1 - 2b_1 + 1)\epsilon + O(\epsilon^2) \equiv 0. \tag{1.50}$$

By the fundamental theorem of perturbation theory,

$$b_0^2 - 2b_0 = 0 \tag{1.51}_0$$

$$2b_0 b_1 - 2b_1 + 1 = 0. \tag{1.51}_1$$

The conjecture (1.46) implies, for the root we seek, that $w(0) \neq 0$. Thus, the solution of $(1.51)_0$ we want is

$$b_0 = 2. \tag{1.52}_0$$

From $(1.51)_1$,

$$b_1 = -1/2. \tag{1.52}_1$$

The root of $R_\epsilon(z)$ we seek therefore has the representation

$$w_2(\epsilon) = 2 - \frac{1}{2}\epsilon + O(\epsilon^2), \tag{1.53}$$

which, as a result of (1.47), produces

$$z_2(\epsilon) = \frac{2}{\epsilon} - \frac{1}{2} + O(\epsilon). \tag{1.54}$$

Thus, we have reproduced (1.21), *without using the quadratic formula*.

Exercise 1.5. Find the first three non-zero terms in the expansions of each of the roots of $\epsilon z^3 - z^2 + 1$.

The analysis of these two simple examples has suggested several basic principles, including the idea of reducing singular problems to regular ones by a change of variable and then seeking solutions of the form (1.22). But are the solutions of every regular problem regular? The answer is "not necessarily," as the following problem shows.

> Find an expansion for the roots of $P_\epsilon(z) = z^2 - 2\epsilon z - \epsilon$,
> useful if $|\epsilon|$ is small.

The problem seems regular since no roots are lost if $\epsilon = 0$. Therefore, we seek solutions of the form (1.22). Following the familiar routine, we substitute (1.22) into $P_\epsilon(z)$, collect coefficients of like powers of ϵ, and require that the resulting expression be identically zero if ϵ is sufficiently small. With, say, $N = 2$, this yields

$$a_0^2 + (2a_0a_1 - 2a_0 - 1)\epsilon \tag{1.55}$$

$$+ (a_1^2 + 2a_0a_2 - 2a_1)\epsilon^2 + O(\epsilon^3) \equiv 0.$$

By the Fundamental Theorem of Perturbation Theory,

$$a_0^2 = 0 \tag{1.56}_0$$

$$2a_0a_1 - 2a_0 - 1 = 0 \tag{1.56}_1$$

$$a_1^2 + 2a_0a_2 - 2a_1 = 0. \tag{1.56}_2$$

The solution of $(1.56)_0$ is $a_0 = 0$ so that $(1.56)_1$ reduces to $-1 = 0$. This contradiction tells us there is no root of $z^2 - 2\epsilon z - \epsilon$ of the form (1.22).

We could fall back on the quadratic formula, but the purpose here is to develop a line of reasoning that does not appeal to formulas for exact solutions. (Why? Because, for arbitrary polynomials of degree higher than four, no such formulas exist.) Instead, we start from the fact that the roots of

$$P_\epsilon(z) = z^2 - 2\epsilon z - \epsilon \qquad (1.57)$$

must approach zero as ϵ does. We next ask how fast does this process occur? Let us assume that as $\epsilon \to 0$,

$$z(\epsilon) \sim \epsilon^p b_0, \ p > 0, \ b_0 \neq 0. \qquad (1.58)$$

The symbol $\sim$ means here that $\lim \epsilon^{-p} z(\epsilon) = b_0$. Of course, there are many other ways that $z(\epsilon)$ *might* approach zero, say $z(\epsilon) \sim \epsilon \ln \epsilon$, but first, we want to consider a simple power dependence.

The assumption (1.58) suggests the change of variable

$$z(\epsilon) = \epsilon^p w(\epsilon) \, , \ w(0) \neq 0. \qquad (1.59)$$

Substituting (1.59) into (1.57), we have

$$P_\epsilon(\epsilon^p w(\epsilon)) = \epsilon^{2p} w^2(\epsilon) - 2\epsilon^{p+1} w(\epsilon) - \epsilon. \qquad (1.60)$$

As $\epsilon \to 0$, the largest of the three terms ϵ^{2p}, ϵ^{p+1}, ϵ^1 is the one with the smallest exponent. If $p > \frac{1}{2}$, then, as is clear from Fig. 1.2, min $\{2p, p+1, 1\} = 1$. Hence, $\epsilon^{-1} P_\epsilon(\epsilon^p w(\epsilon)) \sim -1$. But this is a contradiction since the left side must be identically zero for ϵ sufficiently small. Likewise, we get a contradiction if $p < \frac{1}{2}$ because in this situation min $\{2p, p+1, 1\} = 2p$ which implies that $\epsilon^{-2p} P_\epsilon(\epsilon^p w(\epsilon)) \sim w^2(0) \neq 0$. The only possibility left is $p = \frac{1}{2}$. In this case,

$$\epsilon^{-1} P_\epsilon(\epsilon^{1/2} w(\epsilon)) = w^2(\epsilon) - 2\epsilon^{1/2} w(\epsilon) - 1 \sim w^2(0) - 1, \qquad (1.61)$$

which implies that $w(0) = \pm 1$.

Observe in (1.61) that it is $\epsilon^{1/2}$ that appears and *not* ϵ. This suggests that, in addition to the change of variable (1.59), we introduce the *change of parameter.*

$$\beta = \epsilon^{1/2} \qquad (1.62)$$

Thus, the problem of finding the roots of $z^2 - 2\epsilon z - \epsilon$ for ϵ sufficiently small has been replaced by the equivalent problem of finding the roots of

$$R_\beta(w) \equiv w^2 - 2\beta w - 1 \qquad (1.63)$$

for β sufficiently small. If our analysis has been correct, then $R_\beta(w)$ should have two regular roots of the form

$$w(\beta) = b_0 + b_1 \beta + \cdots + b_N \beta^N + O(\beta^{N+1}). \qquad (1.64)$$

Substitution of (1.64) into (1.63) leads to the sequence of equations

$$b_0^2 - 1 = 0 \tag{1.65}_0$$

$$2b_0b_1 - 2b_0 = 0 \tag{1.65}_1$$

$$b_1^2 + 2b_0b_2 - 2b_1 = 0 \tag{1.65}_2$$

$$\vdots$$

whose solutions are, sequentially,

$$b_0 = \pm 1 \tag{1.66}_0$$

$$b_1 = 1 \tag{1.66}_1$$

$$b_2 = \pm \tfrac{1}{2} \ . \tag{1.66}_2$$

Substituting into (1.64), (1.62), and (1.59), we find that the two roots of our original polynomial $z^2 - 2\epsilon z - \epsilon$ are, with $N = 3$ in (1.64),

$$z_1(\epsilon) = \epsilon^{1/2} + \epsilon + \frac{1}{2}\epsilon^{3/2} + O(\epsilon^2) \tag{1.67}$$

$$z_2(\epsilon) = -\epsilon^{1/2} + \epsilon - \frac{1}{2}\epsilon^{3/2} + O(\epsilon^2) \ . \tag{1.68}$$

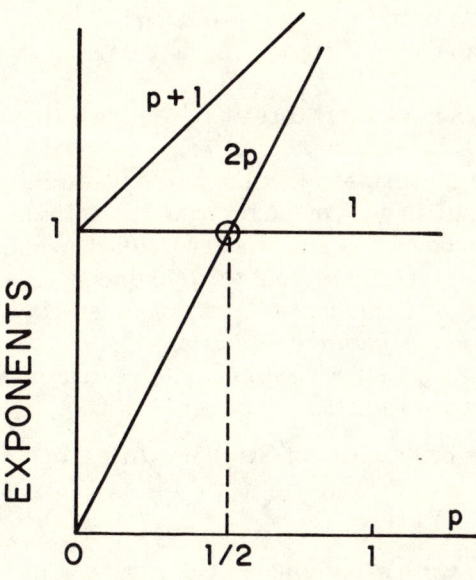

Fig. 1.2. Graphs of the exponents of ϵ in (1.60).
Exponents are equal at intersections.

Exercise 1.6 Verify these results by using the quadratic formula [and show that the error terms are actually smaller than indicated in (1.67) and (1.68)].

Exercise 1.7. Determine explicitly the first two terms in the expansions of the three roots of $z^3 - \epsilon z^2 - \epsilon^2$.

Exercise 1.8. Consider the linear algebraic equations

$$\epsilon^2 x - 2\epsilon y + z = \epsilon$$

$$-x - \epsilon^2 y + \epsilon z = 1$$

$$\epsilon x + \epsilon y + z = \epsilon^2.$$

As $\epsilon \to 0$, $x \sim a_0 \epsilon^p$, $y \sim b_0 \epsilon^q$, $z \sim c_0 \epsilon^r$. Assume regular expansions of the form $\epsilon^{-p} x = a_0 + a_1 \epsilon + \ldots$, etc., and use Cramer's Rule to determine $p, q, r, a_0, b_0, c_0,\ a_1, b_1, c_1$.

Exercise 1.9. Determine the dominant behavior as $\epsilon \to 0$ of the eigenvalues of the coefficient matrix of the system of equations in Exercise 1.8.

Accuracy of a Regular Expansion. Suppose that we can identify the first $N + 1$ terms in (1.22) with the first $N + 1$ terms of a power series. If the series has a radius of convergence $\rho \neq 0$, then, if $|\epsilon| < \rho$, we can compute $z(\epsilon)$ to any degree of precision by making N large enough. Indeed, the examples considered so far have been of this type with $\rho = 1$.

It may happen, however, that $\rho = 0$, in which case $z(\epsilon)$ can be computed from (1.22) *only to within some irreducible error.* That is, for each value of ϵ, it may happen that the remainder term in (1.22) can be made only so small. If we compute terms beyond a certain point, the approximation becomes less and less accurate. How many terms should be kept? With no information about the remainder term except that it is $O(\epsilon^{N+1})$, the rule of thumb is: *stop computing terms when the $(n+1)$st term in (1.22) is larger than the nth term.*

The following is a classic example of the situation described above, but may be skipped with no loss of continuity.

The Small ϵ Expansion of Stieltjes' Integral

$$I(\epsilon) = \int_0^\infty \frac{e^{-t}\, dt}{1 + \epsilon^2 t} \tag{1.69}$$

This shouldn't be too hard to find as $I(0) = \int_0^\infty e^{-t}\, dt = 1$. To account for the effect of ϵ in (1.69), ϵ must be put into the numerator of the integrand.

The following formula for a geometric series with a remainder is of great use.

$$\frac{1}{1-x} = 1 + x + x^2 + \cdots + x^n + \frac{x^{n+1}}{1-x} \ , \ x \neq 1. \ (1.70)$$

Exercise 1.10. Verify (1.70).

Setting $x = -\epsilon^2 t$ and substituting (1.70) into (1.69), we have

$$I(\epsilon) = \int_0^\infty e^{-t}[1 - \epsilon^2 t + \cdots + (-1)^n \epsilon^{2n} t^n]dt \qquad (1.71)$$
$$+ (-1)^{n+1}\epsilon^{2n+2}I_{n+1}(\epsilon),$$

where

$$I_{n+1}(\epsilon) = \int_0^\infty \frac{e^{-t} \ t^{n+1}dt}{1 + \epsilon^2 t}. \qquad (1.72)$$

Each of the integrals in the first line of (1.71) can be evaluated by techniques of first-year calculus:

$$\int_0^\infty e^{-t}t^n dt = n! \qquad (1.73)$$

Thus

$$I(\epsilon) = 1 - \epsilon^2 + 2!\epsilon^4 - \cdots + (-1)^n n!\epsilon^{2n} + (-1)^{n+1}\epsilon^{2n+2}I_{n+1}(\epsilon)$$
$$\equiv \Sigma_0^n(-1)^k \ k!\epsilon^{2k} + R_{n+1}(\epsilon). \qquad (1.74)$$

This expansion is of the same form as (1.22) with ϵ^2 in place of ϵ and $N = 2n$. To show that it is regular, we need only bound $I_{n+1}(\epsilon)$ by a constant independent of ϵ. But $1 + \epsilon^2 t \geq 1$ if $t \geq 0$. Hence,

$$|I_{n+1}(\epsilon)| = \int_0^\infty \frac{e^{-t}t^{n+1}dt}{1 + \epsilon^2 t} \leq \int_0^\infty e^{-t}t^{n+1}dt = (n+1)! \ , \ Q.E.D. \quad (1.75)$$

To show that $\Sigma_0^n(-1)^k \ k!\epsilon^{2k}$ *cannot* approximate $I(\epsilon)$ to arbitrary precision, *no matter how we choose n*, we use the last link in the following chain of lower bounds on $I_{n+1}(\epsilon)$:

$$|I_{n+1}(\epsilon)| > \int_1^\infty \frac{e^{-t}t^{n+1}dt}{1+\epsilon^2 t} \qquad (1.76)$$

$$> \int_1^\infty \frac{e^{-t}t^{n+1}dt}{t+\epsilon^2 t}$$

$$= \frac{e^{-1}}{1+\epsilon^2}\int_0^\infty e^{-s}(1+s)^n \ ds \ , \ t = 1+s$$

$$> \frac{e^{-1}}{1+\epsilon^2} \int_0^\infty e^{-s} s^n \, ds$$

$$= \frac{e^{-1}n!}{1+\epsilon^2}$$

With this result and (1.75), we obtain from (1.74),

$$L_{n+1}(\epsilon) \equiv \frac{\epsilon^{2n+2}n!}{e(1+\epsilon^2)} < |I(\epsilon) - \Sigma_0^n(-1)^k \, k!\epsilon^{2k}| \tag{1.77}$$

$$< \epsilon^{2n+2}(n+1)! \equiv U_{n+1}(\epsilon).$$

Since

$$U_n(\epsilon) = \epsilon^{2n}n! = [\epsilon^2(n)] \, [\epsilon^2(n-1)]\ldots[\epsilon^2(2)] \, [\,\epsilon^2(1)], \tag{1.78}$$

$\epsilon^{2n} \, n!$ begins to grow once $\epsilon^2 n$—the largest factor on the right side of (1.78)—exceeds one. Therefore given ϵ, $\epsilon^{2n} \, n!$ will be smallest when n is the largest integer n_* such that $\epsilon^2 n_* \leqslant 1$. Moreover, as the right side of (1.78) is the product of n decreasing factors, its value is greater than the nth power of the smallest factor: that is, min $\epsilon^{2n} n! > (\epsilon^2)^{n*} \geqslant \epsilon^{2\epsilon-2}$ if $|\epsilon| < 1$. It now follows from the left side of (1.77) that

$$L_{n+1}(\epsilon) > \frac{\epsilon^{2(1+\epsilon^{-2})}}{e(1+\epsilon^2)}. \tag{1.79}$$

Thus, we make an error at least as large as the right side of (1.79) when we approximate $I(\epsilon)$ by $\Sigma_0^n(-1)^k \, k!\epsilon^{2k}$.

In practice, we would also like to know how small we can make $U_{n+1}(\epsilon)$. It follows from (1.78) that, given ϵ, $U_{n+1}(\epsilon)$ is minimized by taking n as the largest integer such that $n+1 \leqslant \epsilon^{-2}$. For example, if $\epsilon = .5$, take $n = 3$. This guarantees that I(.5) is approximated by $1 - (.5)^2 + 2(.5)^4 - 6(.5)^6 = .78125$ to within an absolute error that is no greater than $(.5)^8 4! = .09375$. (The correct value of I(.5) is $4e^4 E_1(4) = .82538\ldots$, where E_1 is the exponential integral.)

Exercise 1.11. Graph the upper bound on the error in the optimum approximation to $I(\epsilon)$ for $0 \leqslant \epsilon \leqslant 1$.

Exercise 1.12. Given ϵ, use the double inequality[4]

$$\sqrt{2\pi n}\ ^{n+1/2}\exp[-n + (12n+1)^{-1}] < n! < \sqrt{2\pi n}\ ^{n+1/2}\exp[-n + (12n)^{-1}] \qquad (1.80)$$

to obtain lower and upper bounds on $L_{n+1}(\epsilon)$ and $U_{n+1}(\epsilon)$, respectively, in terms of ϵ only. According to these estimates, if $\epsilon = .1$ then $\Sigma_0^{99}(-1)^k$ $k!(.01)^k$ should approximate $I(.1)$ to within an error of $O(10^{-43})$.

Asymptotic Sequences. As we have tried to show by simple examples, the goal of perturbation theory is to reduce singular problems to regular ones. These we try to solve approximately by solving a sequence of simpler problems that have been purged of all small parameters. Working different problems has forced us to keep enlarging our concept of a regular expansion (solution). At first, a regular expansion was merely a power series in ϵ with a *known* radius of convergence. Then, it was a power series convergent for some value of ϵ *sufficiently small*. Next we decided that a *finite power series with a remainder estimate* might be a better definition of a regular expansion. When we saw that an equation containing integral powers of ϵ could have a solution involving fractional powers of ϵ, we again broadened our definition.

Our final definition is this: *a regular expansion is an expression of the form*

$$z(\epsilon) = a_0\phi_0(\epsilon) + a_1\phi_1(\epsilon) + \cdots + a_n\phi_n(\epsilon) + R_{n+1}(\epsilon),$$
$$R_{n+1}(\epsilon) = O(\phi_{n+1}(\epsilon)), \qquad (1.81)$$

where

$$\lim \phi_{n+1}(\epsilon)/\phi_n(\epsilon) = 0 \text{ as } \epsilon \to 0. \qquad (1.82)$$

Any sequence of functions having the property (1.82) is said to be an *asymptotic sequence*. You should convince yourself that the Fundamental Theorem of Perturbation Theory, stated in the box containing (1.33) and (1.34), continues to hold if ϵ^k is everywhere replaced by $\phi_k(\epsilon)$.

For the simple problems considered in the rest of the book, expansions of the form (1.22) are sufficient, but it is important to note that expansions of the form (1.81) are not uncommon. For example, suppose that we need the small ϵ expansion of

$$f(\epsilon) = e^\epsilon + \epsilon^{1+\epsilon}. \qquad (1.83)$$

From the power series expansion for the exponential function, we have

---

[4]See Feller, *An Introduction to Probability and Its Applications*, Vol. 1, 3rd Ed., Wiley, 1968, p. 54.

$$e^\epsilon = 1 + \epsilon + \tfrac{1}{2}\epsilon^2 + \cdots \tag{1.84}$$

$$\epsilon^{1+\epsilon} = \epsilon e^{\epsilon ln\epsilon} \tag{1.85}$$

$$= \epsilon(1 + \epsilon ln\epsilon + \tfrac{1}{2}\epsilon^2 ln^2\epsilon + \ldots).$$

Thus

$$f(\epsilon) = 1 + 2\epsilon + \epsilon^2 ln\epsilon + \tfrac{1}{2}\epsilon^2 + \tfrac{1}{2}\epsilon^3 ln^2\epsilon + \cdots$$

$$\equiv \phi_0(\epsilon) + 2\phi_1(\epsilon) + \phi_2(\epsilon) + \tfrac{1}{2}\phi_3(\epsilon) + \tfrac{1}{2}\phi_4(\epsilon) + \cdots. \tag{1.86}$$

Clearly, the ϕ_n's satisfy (1.82).

Differential Equations. Though finding the roots of a polynomial or the value of an integral is necessary in the analysis of a variety of problems, it is rare that a polynomial or integral itself models a phenomenon. (We often try to fit experimental data with polynomials, but this is not true modelling. Rather, it is an admission that the mechanism producing the data is either too obscure or too complicated to have predictive value). More often, differential equations (DE's) are used as models, and for many phenomena, such as the motion of a point-mass in a gravitational field, the kinetics of a chemical reaction, or the growth of a colony of bacteria, *ordinary* DE's (ODE's) suffice. Our main concern shall be to determine when and how such ODE's can be solved *approximately* by perturbation methods.

Linear, constant coefficient ODE's are encountered everywhere. In theory, they are easy to solve. It is important to remember that if the independent variable is, say t, then the first step toward a solution is to look for solutions of the form e^{zt}, where z is a constant to be determined. In the case of a single equation of degree n (or a system of n first-order equations), this leads to finding the roots of a polynomial in z of degree n. If the coefficients of the DE (or DE's) contain a parameter ϵ, so will the polynomial, and if ϵ is small, we may construct close approximations to the roots by using some of the techniques mentioned above. These techniques will be developed systematically in Chapter II.

Special Features of ODE's. The solutions of ODE's are subject to *auxiliary conditions*, usually in the form of *initial conditions* or *boundary conditions*, which may depend on the parameter ϵ. More importantly, if the roots of the associated polynomial of a constant coefficient ODE depend on ϵ, then the solution will depend on *two* variables, ϵ and t. *The interplay of these is crucial.* A small disturbance, as reflected by the presence of a parameter ϵ in a DE or its auxiliary conditions, may lead to a large effect because, when the range of t is unbounded, terms such as ϵt can grow large (singularity in the domain), or, if the range of t is bounded, say $0 \leq t \leq 1$, terms such as t/ϵ can grow large (singularity in the model). These phenomena are examples of *nonuniformities*.

Nondimensionalization of the equations describing a physical system is done by expressing each variable having physical units as a fraction of some fixed *intrinsic* quantity. Nondimensionalization is an extremely useful procedure, even when a problem is not amenable to perturbation methods.

From physics we know that the period of a simple, frictionless pendulum of length l, oscillating with an arbitrarily small angular amplitude, is $2\pi(l/g)^{1/2}$, where g is the strength of the gravitational field. Thus, in studying large oscillations, it is natural to measure the actual time as a fraction of $(l/g)^{1/2}$, especially if we are interested in motions in which the amplitude and friction are small. The fraction itself is called the *dimensionless time*. As another example, the equation of motion of a tightly-stretched string takes a particularly concise form if distance along the string is represented by xL and time by $tL(\rho/T)^{1/2}$. Here L is the length of the string, ρ is its mass per unit length, assumed constant, and T is the tension. Physically, $L(\rho/T)^{1/2}$ is the time it takes a small disturbance to move from one end of the string to the other.

Often there may be more than one way to nondimensionalize the equations of a mathematical model. Consider the equations describing the fall of a ball released from rest at a distance h above the ground. Assuming a constant gravitational field of strength g and an atmosphere of constant density ρ, we could express the velocity as a fraction of $(2gh)^{1/2}$, the velocity of impact in vacuum. On the other hand, we could express the velocity as vV, where V, determined by equating the drag on the ball to the gravitational force, is the terminal velocity the ball would approach if released from an infinite height. If the influence of the drag on the impact velocity is *expected* to be small, it is reasonable to choose $(2gh)^{1/2}$ as the unit of velocity; otherwise V is perhaps the best choice.

A great virtue of nondimensionalization is that it makes it possible to represent an *infinity* of physical problems by a *single* mathematical problem. For example, the equation for the angular motion, $\theta(t)$, of a simple frictionless pendulum of any length l in a gravitational field g, released from rest at an angle ϵ, is

$$\ddot{\theta} + \sin \theta = 0 \, , \, t > 0 \, , \, \theta(0) = \epsilon \, , \, \dot{\theta}(0) = 0. \qquad (1.87)$$

Here, t is the dimensionless time discussed earlier and $\dot{\theta} = d\theta/dt$.

Exercise 1.13. Derive the *DE* in (1.87).

Because of conservation of energy, the pendulum will return to its initial angular position after some dimensionless period of time t_*, which depends on ϵ. The period of oscillation of the physical pendulum is $(l/g)^{1/2}t_*$. Thus, for a given initial angular displacement ϵ, computation of the single number t_* gives the period of a physical pendulum of *any* length l swinging in a gravitational field of *any* strength g.[5]

Even if effects such as friction are included, a single solution of the dimensionless equation of motion still represents an infinite number of physical solutions. For instance, suppose that our pendulum oscillates in a vacuum, but that its pivot exerts a torque $\eta\dot{\theta}$ opposing the motion. Here η is a constant with the units of $(mass) \times (length)^2/(time)$. In place of (1.87), we have

$$\ddot{\theta} + 2\alpha\dot{\theta} + \sin\theta = 0 \,, \ t > 0 \,, \ \theta(0) = \epsilon \,, \ \dot{\theta}(0) = 0, \qquad (1.88)$$

where $2\alpha = \eta/(mg^{1/2}l^{3/2})$ and m is the mass of the bob of the pendulum. The period of motion is no longer constant. However, for *given* values of ϵ and α, (1.88) represents an *infinite* class of *equivalent* physical problems—two physical problems being equivalent if they have equal values of α and ϵ.

Exercise 1.14. An elastic beam of section modulus EI, resting on an elastic foundation of modulus k, is under a tension T and a distributed downward force/length $p(s)$, where s is distance along the beam measured from some convenient point. The small vertical deflection w of the beam satisfies the DE

$$\frac{d^2}{ds^2} \left(EI \, \frac{d^2w}{ds^2} \right) - T \, \frac{d^2w}{ds^2} + kw = p. \qquad (1.89)$$

For simplicity, assume that EI, T, and k are constants, expressed in some common set of physical units. Show, by setting $s = Lx$ and appropriately choosing L (which has units of length), that (1.89) can be given the dimen-

---

[5]Actually, from the principles of dimensionless analysis alone, it may be concluded that the period of a simple frictionless pendulum must be of the form $(l/g)^{1/2}t_*(\epsilon)$. The function $t_*(\epsilon)$, however, can be determined only by an integration of (1.87), which we shall do in Chapter IV. A more striking example of the power of dimensional analysis comes from subsonic aerodynamics where it is shown that, when viscosity is negligible, the drag on an object immersed in a uniform flow of infinite extent is given by $C_D \rho L^2 V^2$, where ρ is the air density, L is some characteristic dimension of the object, and V is the velocity of the flow far from the object. The dimensionless drag coefficient C_D depends only on the shape, of the object. C_D is extremely difficult to compute from the equations of fluid dynamics, even for the simplest shapes, and is therefore usually determined experimentally from wind tunnel tests. See Rauscher, M., *Introduction to Aeronautical Dynamics*, Wiley, 1953, Sections 10.7 & 10.8.

sionless form

$$\epsilon^2 y'''' - y'' + y = \beta f(x), \tag{1.90}$$

where $y' = dy/dx$.

Another useful feature of nondimensionalization is that it sometimes reveals disparate physical systems to be identical mathematically. (This opens the possibility of solving physical problems by analogy.) Consider the forced, damped, linear spring-mass system in Fig. 1.3. Its motion is described by the differential equation and initial conditions

$$m \frac{d^2\xi}{d\tau^2} + c \frac{d\xi}{d\tau} + k\xi = F(\tau), \tau > 0$$

$$\xi(0) = \xi_0, \dot{\xi}(0) = 0. \tag{1.91}$$

(Other initial conditions could have been chosen). Here ξ is the location of the mass measured from the rest position of the spring, τ is the time, m is the mass, c is a damping coefficient, and k the spring constant. The units of $\{\xi, \tau, m, c, k\}$ are, say, {centimeters, seconds, grams, gram-centimeters per second, gram-centimeters per second squared}.

The study of differential equations reveals that homogeneous solutions of the *DE* in (1.91) in the absence of damping ($c = 0$) are of the form $\cos(\tau\sqrt{k/m})$, $\sin(\tau\sqrt{k/m})$. Thus, the period of the free oscillations of

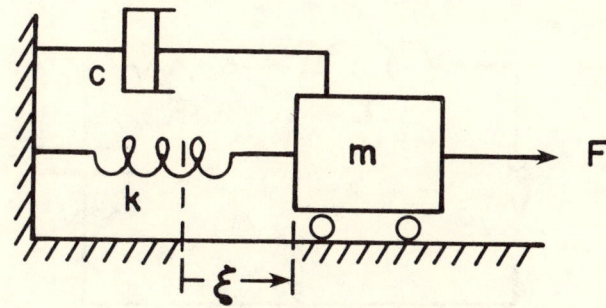

Fig. 1.3. A linear, damped, spring-mass system.

the linear spring-mass system is $2\pi\sqrt{m/k}$. Suppose that we are interested in weak damping forces. Because these will little change the undamped period of the system, it is natural to introduce the dimensionless time

$$t = \tau\sqrt{k/m}. \qquad (1.92)$$

Furthermore, in the absence of a forcing term, damping will cause the system to dissipate energy continually. In view of the initial conditions, the maximum displacement will never exceed ξ_0 in magnitude. This suggests that we introduce the dimensionless displacement

$$x = \xi/\xi_0. \qquad (1.93)$$

(Different initial conditions would suggest a different way of nondimensionalizing ξ). With the change of variables (1.92) and (1.93), (1.91) is transformed into

$$\ddot{x} + 2\epsilon\dot{x} + x = f(t)\,, 0 < t$$

$$\qquad (1.94)$$

$$x(0) = 1\,, \dot{x}(0) = 0,$$

where

$$2\epsilon = c\,/\,\sqrt{mk}\,. \qquad (1.95)$$

Now consider the closed circuit sketched in Fig. 1.4. that consists of a linear inductor, a linear resistor, and a linear capacitor in series and

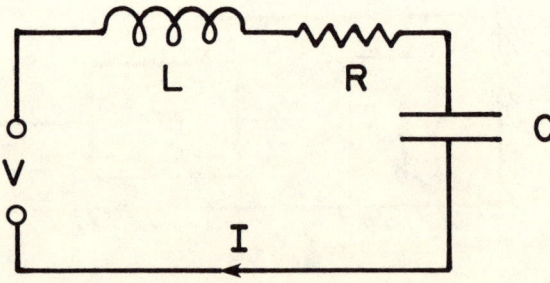

Fig. 1.4. A linear LRC circuit.

driven by an applied, time-varying voltage $V(T)$—a so-called LRC circuit. The current I in the circuit satisfied the differential equation

$$L\frac{d^2I}{dT^2} + R\frac{dI}{dT} + \frac{I}{C} = \frac{dV}{dT}, \, 0 < T \qquad (1.96)_1$$

and a set of initial conditions, which we take to be

$$I(0) = I_0, \, \dot{I}(0) = 0. \qquad (1.96)_2$$

Here L, R, and C are, respectively, the inductance, resistance, and the capacitance of the system, measured, say, in units of Henrys, Ohms, and Farads, respectively.

It is now obvious, with the change of variables

$$t = T / \sqrt{LC}, \, x = I / I_0 \qquad (1.97)$$

and the introduction of the dimensionless parameter and function

$$2\epsilon = R\sqrt{C/L}, \, f \equiv \frac{C}{I_0}\frac{dV}{dT}, \qquad (1.98)$$

that (1.96) reduces precisely to (1.94). That is, the forced motion of a damped linear oscillator is identical, mathematically, to the current in a driven LRC circuit. Find the solution for one of these systems and the solution of the other comes free.

Exercise 1.15. As indicated in Fig. 1.5. the outer edge of an annular membrane is attached to a rigid plate with a hole of radius b. The inner edge of the membrane is attached to a movable rigid disk of radius a which is under a downward vertical force P. The membrane is of constant thickness h and obeys a linear, isotropic stress-strain law. According to Föppl's theory, the equilibrium and compatibility conditions for the membrane (assuming that fibers undergo small but finite angles of rotation) can be reduced to the single equation

$$\frac{d}{dr}\left[r\frac{d}{dr}(rT)\right] - T = -\frac{1}{2}\frac{EhP^2}{r^2T^2}, \qquad (1.99)$$

where T is the radial tension and E is Young's modulus, an elastic parameter, here taken constant.

a. Show with the change of variables

$$\xi = r^2, \, f = \xi T, \qquad (1.100)$$

that (1.99) takes the form

$$\frac{d^2 f}{d\xi^2} = -\frac{1}{8}\frac{EhP^2}{f^2} \,.$$

(1.101)

b. Set

$$\xi = b^2 x \,,\; f = \lambda y$$

(1.102)

and determine the constant λ so that (1.101) takes the dimensionless form

$$y'' = -\frac{1}{y^2} \,.$$

(1.103)

c. State Newton's Law for a mass-point falling towards the center of attraction of an inverse square gravitational force. Nondimensionalize the equation so that it reduces to (1.103).

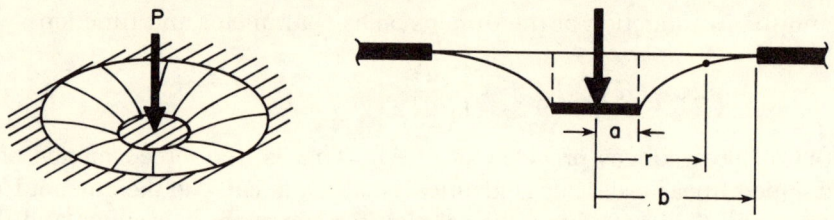

Fig. 1.5. An initially slack membrane under a central load.

CHAPTER II:
ROOTS OF POLYNOMIALS

In Chapter I we looked at several simple polynomials with coefficients depending on a parameter ϵ. In each case we found that constructing a useful expansion for the roots reduced, ultimately, to computing, one-by-one, the coefficients in a regular expansion of the form (1.22). In some cases, all of the roots were regular. In other cases, a change of variable and perhaps parameter was necessary to get an associated polynomial that *did have* regular roots.

In this chapter we shall develop a procedure for representing the roots of a polynomial of the form

$$P_\epsilon(z) = (1 + b_0\epsilon + c_0\epsilon^2 + \cdots) + A_1\epsilon^{\alpha_1}(1 + b_1\epsilon + \cdots)z$$

$$+ \cdots + A_n\epsilon^{\alpha_n}(1 + b_n\epsilon + \cdots)z^n, \tag{2.1}$$

where the α_i's are rational numbers and the b_i's, c_i's, $\cdots$ are constants. Further, each of the factors $1 + b_k \epsilon + \cdots$, $k = 0, 1, \ldots, n$, is assumed to be regular, *i.e.*, to have an expansion of the form (1.22). The polynomials mentioned in Chapter I are special cases of (2.1).

Exercise 2.1. Replace each of the following polynomials by a polynomial of the form (2.1) having the same non-zero roots:

(a) $\epsilon - 2z + z^2$

(b) $2 + \epsilon z + \epsilon^{-1}z^4$

(c) $\epsilon z - z^3$.

As $\epsilon \to 0$, the different roots of $P_\epsilon(z)$ may approach zero, finite non-zero values, or infinity. For example, $1 + \epsilon^{-1}z + \epsilon^{-1}z^2 + z^3$ has roots of each type, as you'll be asked to show later in an exercise. In Chapter I we studied the roots of $\epsilon z^2 - 2z + 1$ and $z^2 - 2\epsilon z - \epsilon$. We found that the first had a non-regular root $z_2(\epsilon) \sim 2\epsilon^{-1}$ and that the second had two non-regular roots, $z_1(\epsilon) \sim \epsilon^{1/2}$ and $z_2(\epsilon) \sim -\epsilon^{1/2}$. Expansions for these non-regular roots were obtained by introducing the change of variable $z(\epsilon) = \epsilon^p w(\epsilon)$ and then determining p by requiring that $w(0) \neq 0$. This is the key to finding useful expansions for the roots of (2.1) and motivates the following

THEOREM: Each root of (2.1) is of the form

$$z(\epsilon) = \epsilon^p w(\epsilon), \ w(0) \neq 0 , \tag{2.2}$$

where $w(\epsilon)$ is a continuous function of ϵ for ϵ sufficiently small.

PROOF: First, we shall determine an algorithm for finding what we shall call *the proper values* of p. Then we shall establish (2.2) and the continuity of $w(\epsilon)$.

If (2.2) is to hold, then, from (2.1),

$$P_\epsilon(\epsilon^p w) = Q_\epsilon(w) + \epsilon(b_0 + \epsilon^{\alpha_1 + p} b_1 A_1 w + \cdots + \epsilon^{\alpha_n + np} b_n A_n w^n) + \cdots , \tag{2.3}$$

where

$$Q_\epsilon(w) \equiv 1 + \epsilon^{\alpha_1 + p} A_1 w + \cdots + \epsilon^{\alpha_n + np} A_n w^n). \tag{2.4}$$

Let $z(\epsilon) = \epsilon^p w(\epsilon)$ be a root of $P_\epsilon(z)$, i.e., $P_\epsilon(\epsilon^p w(\epsilon)) \equiv 0$ for all ϵ. As $Q_\epsilon(w)$ is a continuous function of w and also of ϵ if $\epsilon \neq 0$, it follows that if $w(\epsilon)$ is continuous, then

$$\lim Q_\epsilon(w(0)) = 0. \tag{2.5}$$

In view of (2.4), (2.5) implies that if $w(0) \neq 0$ then *at least two* of the exponents in the set

$$E = \{0, \ \alpha_1 + p, \ \cdots , \alpha_n + np\} \tag{2.6}$$

must have *identical, minimal values*. The reason is clear: If $w(0) \neq 0$, then as $\epsilon \to 0$ the dominant terms in $Q_\epsilon(w)$ are those whose ϵ – factors have the smallest exponents. If there were but one minimal exponent, say $\alpha_k + kp$, then, since $P_\epsilon(\epsilon^p w(\epsilon)) \equiv 0$, it would follow from (2.3) that $\lim \epsilon^{-(\alpha_k + kp)} P_\epsilon(\epsilon^p w(\epsilon)) = A_k w^k(0) = 0$. But this is a contradiction, because if $\alpha_k + kp$ belongs to the set (2.6), $A_k \neq 0$.

To select the proper values of p, look at Fig. 2.1, the linear graphs of the exponents in (2.6). The circled intersections correspond to value of p where two or more exponents have equal, minimal values. If p is sufficiently large, the smallest exponent in the set (2.6) will be 0. As p decreases (imagine a vertical line in Fig. 2.1 moving from right to left), there will be a first, largest proper value p_1 such that at least two exponents in (2.6) have the value $e_1 = 0$. One (and only one) of the graphs passing through the point (p_1, e_1) will have a largest slope n_1. Now continue to decrease p until a second proper value p_2 is reached for which at least two of the exponents in (2.6) take on some identical, minimal value e_2. The slopes of the graphs through (p_2, e_2) range from a minimum,

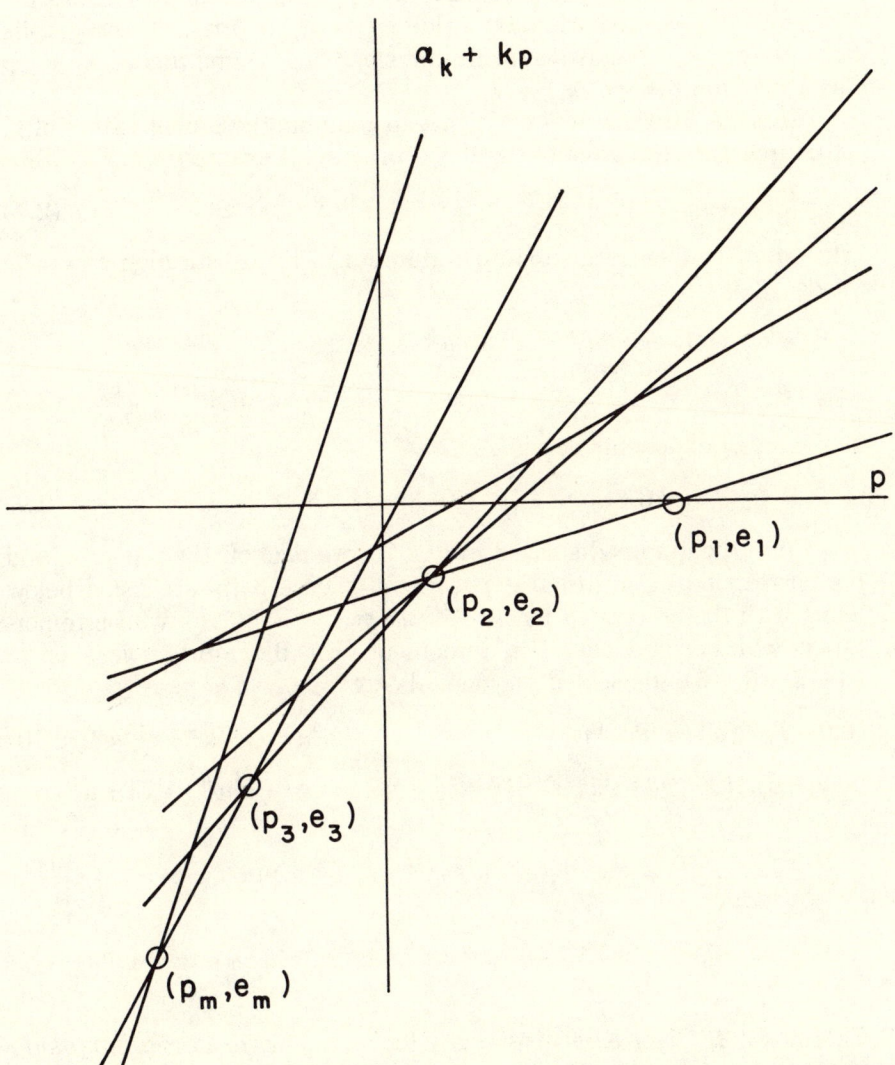

Fig. 2.1. Graphs of the exponents of ϵ in equation (2.4).

n_1, to some maximum, n_2. Proceed in this fashion until the last and smallest proper value, p_m, is reached. For $p = p_m$ at least two exponents in (2.6) will have the common value $e_m = \alpha_n + np_m$. Of the graphs through (p_m, e_m), one has a minimum slope, n_{m-1}, and that of $\alpha_n + np$ has a maximum slope, $n_m = n$.

To make sure that it is clear how to compute the proper values of p, let us interrupt the proof of the theorem with an example,

$$P_\epsilon(z) = 1 + z^3 + \epsilon^6 z^6 + 2\epsilon^9 z^6 + \epsilon^{12} z^8 + \epsilon^{18} z^9 \tag{2.7}$$

(rigged, of course, to give simple answers.) The substitution $z = \epsilon^p w$ yields

$$P_\epsilon(\epsilon^p w) = 1 + \epsilon^{3p} w^3 + \epsilon^{6+6p} w^6 + 2\epsilon^{9+7p} w^7 + \epsilon^{12+8p} w^8$$

$$+ \epsilon^{18+9p} w^9, \tag{2.8}$$

so the set of exponents of ϵ is

$$E = \{0, 3p, 6+6p, 9+7p, 12+8p, 18+9p\}. \tag{2.9}$$

From the graphs displayed in Fig. 2.2 we read off the proper p_j and the corresponding minimal exponents e_j. These pairs are listed below along with the associated forms $T_\epsilon^{(j)}(w) \equiv \epsilon^{-e_j} P_\epsilon(\epsilon^{p_j} w)$, whose importance will become clear in a moment. For the four choices of p, $(0, -2, -3, -6)$, the scaled polynomials are:

$(0,0)$: $T_\epsilon^{(1)}(w) \equiv \epsilon^0 P_\epsilon(\epsilon^0 w) = 1 + w^3 + \epsilon^6 w^6 + 2\epsilon^9 w^7 + \epsilon^{12} w^8 + \epsilon^{18} w^9$ (2.10)

$(-2, -6)$: $T_\epsilon^{(2)}(w) \equiv \epsilon^6 P_\epsilon(\epsilon^{-2} w) = w^3 + w^6 + 2\epsilon w^7 + \epsilon^2 w^8 + \epsilon^6(1 + w^9)$

$$\tag{2.11}$$

$(-3, -12)$: $T_\epsilon^{(3)}(w) \equiv \epsilon^{12} P_\epsilon(\epsilon^{-3} w) = w^6 + 2w^7 + w^8 + \epsilon^3(w^3 + w^9) + \epsilon^{12}$

$$\tag{2.12}$$

$(-6, -36)$: $T_\epsilon^{(4)}(w) \equiv \epsilon^{36} P_\epsilon(\epsilon^{-6} w) = w^8 + w^9 + 2\epsilon^3 w^7 + \epsilon^6 w^6 + \epsilon^{18} w + \epsilon^{36}$

$$\tag{2.13}$$

Exercise 2.2. Given a set of rationals, $\{0, \cdots, \alpha_n\}$, write a program (using your favorite language or pseudo language) to compute the associated proper values of p.

Back to the proof of the theorem. With the aid of (2.2), we rewrite (2.1) in the form

$$T_\epsilon^{(j)}(w) \equiv \epsilon^{-e_j} P_\epsilon(\epsilon^{p_j} w) = T_0^{(j)}(w) + E_\epsilon^{(j)}(w), \tag{2.14}$$

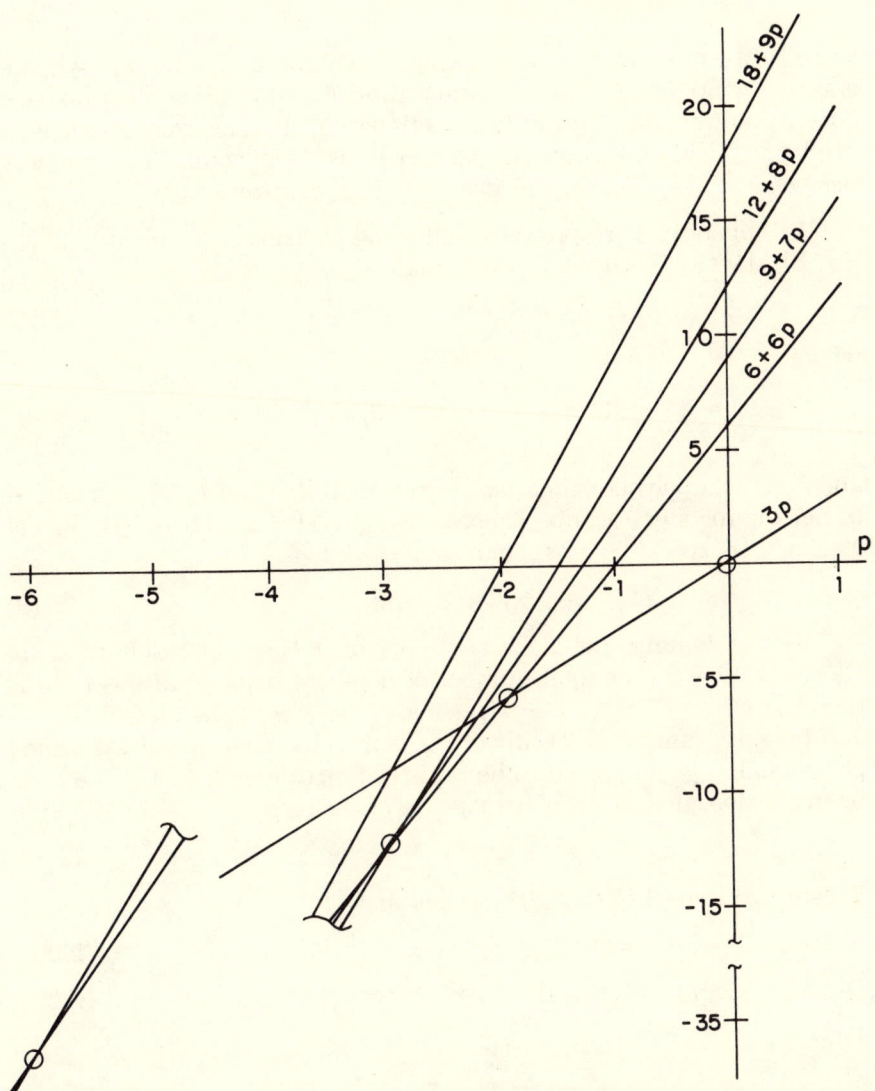

Fig. 2.2. Graphs of the exponents of ϵ in equation (2.8).

where

$$T_0^{(j)}(w) = A_{n_j}(w^{n_j} + \cdots + B_j w^{nn-k}) \tag{2.15}$$

and $E_0^{(j)}(w) = 0$. That is, $T_0^{(j)}(w)$ has no ϵ's and $E^{(j)}(w)$ is zero when ϵ is zero. Equations (2.10)–(2.13) are arranged in this form. What we have done in multiplying P_ϵ by ϵ^{-e_j} and changing variables from z to w is to display explicitly the dominant part of P_ϵ as a polynomial $T_0^{(j)}$ which is *independent of* ϵ. Clearly, $T_0^{(j)}$ has $n_j - n_{j-k}$ non-zero roots.

In Appendix A we prove the following plausible
LEMMA: Consider the polynomial

$$T_\epsilon(z) = T_0(z) + E_\epsilon(z), \tag{2.16}$$

where

$$T_0(z) = z^n + B_1 z^{n-n_1} + \cdots + B_k z^{n-n_k}, \; 0 < n_1 < \cdots n_k \leqslant n \tag{2.17}$$

and $E_\epsilon(z)$ is a polynomial of *any* degree such that $\lim E_\epsilon(z) = 0$ as $\epsilon \to 0$. Let the roots of $T_0(z)$ be denoted by $z_1, z_2, \cdots, z_n$. Then $T_\epsilon(z)$ has at least n roots $z_1(\epsilon), \cdots, z_n(\epsilon)$ such that

$$\lim z_k(\epsilon) = z_k , \; k = 1, 2, \ldots, n . \tag{2.18}$$

By the lemma, the non-zero roots of $T_\epsilon^{(j)}(w)$ approach those of $T_0^{(j)}(w)$ as $\epsilon \to 0$. The total number of non-zero roots of all the $T_0^{(j)}$'s is therefore $(n_1 - 0) + (n_2 - n_1) + \cdots + (n - n_{m-1}) = n$.Q.E.D.

In our example (2.7), all of the non-zero roots of the associated polynomials $T_\epsilon^{(j)}(w)$ appear to be regular. Starting with (2.10), we note, by inspection, that its regular roots are of the form

$$w = a_0 + a_6 \epsilon^6 + O(\epsilon^7). \tag{2.19}$$

Substituted into (2.10), (2.19) implies that

$$1 + a_0^3 = 0 \tag{2.20}_0$$

$$3a_0^2 a_6 + a_0^6 = 0. \tag{2.20}_6$$

Thus

$$a_0 = (-1)^{1/3} = \{e^{i\pi/3}, e^{i\pi}, e^{i5\pi/3}\} \tag{2.21}_0$$

$$a_6 = -a_0^4/3 = a_0/3. \tag{2.21}_6$$

The first three roots of (2.7) are therefore of the form

$$z_k(\epsilon) = \epsilon^0 w_k(\epsilon) = e^{(2k-1)i\pi/3}[1 + \epsilon^6/3 + O(\epsilon^7)] , \; k = 1, 2, 3. \tag{2.22}$$

The regular, non-zero roots of (2.11) are, by inspection, of the form

$$w = b_0 + b_1\epsilon + O(\epsilon^2). \tag{2.23}$$

This expansion, substituted into (2.11), implies that

$$b_0^3 + b_0^6 = 0 \tag{2.24}_0$$

$$3b_0^2 b_1 + 6b_0^5 b_1 + 2b_0^7 = 0. \tag{2.24}_1$$

The non-zero roots of $(2.24)_0$ are, again, the cube roots of -1:

$$b_0 = (-1)^{1/3} = \{e^{i\pi/3}, e^{i\pi}, e^{i5\pi/3}\}. \tag{2.25}_0$$

With $b_0^3 = -1$, $(2.24)_1$ yields

$$b_1 = -(2/3)b_0^2. \tag{2.25}_1$$

Thus the next three roots of (2.7) are of the form

$$z_k(\epsilon) = \epsilon^{-2} w_k(\epsilon) = \epsilon^{-2} e^{(2k-7)i\pi/3}[1 - (2/3)\epsilon e^{(2k-7)i\pi/3} \\ + O(\epsilon^2)]\,,\ k = 4,5,6. \tag{2.26}$$

Equation (2.12) was the result of the intersection of three lines on the graph (see Fig. 2.2). The number of terms in $T_0^{(j)}(w)$ is equal to the number of intersecting lines. When more than two lines intersect, we must consider the possibility of $T_0^{(j)}(w)$ having multiple roots. If multiple roots occur, as happens in (2.12), then the order of the correction terms is somewhat harder to deduce.

Exercise 2.3. By finding exact solutions, deduce the form of the perturbation expansions for roots of

a) $z^2 + 4z - 5 - \epsilon$

b) $z^2 + 2z + 1 - \epsilon$

What interesting differences do you see in the two cases.

Exercise 2.4. Compute the dominant term and one correction term for (2.12) and (2.13). You will have to figure the appropriate order of the correction term based on experience gained in Exercise 2.3.

In general, the non-zero roots of the polynomials $T_\epsilon^{(j)}(w)$ defined by (2.14) need not be regular: the α's in (2.3) and the associated proper values and exponents, (p_j, e_j), may be non-integer rationals. Thus to obtain regular expansion, new parameters must be introduced.

Let

$$\epsilon = \beta^{qj}, \tag{2.29}$$

where q_j is the least common denominator of the set of exponents $\{0, \ldots, \alpha_n + np_j\}$. Then from (2.14)

$$R_\beta^{(j)}(w) \equiv T^{(j)}(w;\beta^{q_j}) = \beta^{-q_j e_j} P(\beta^{q_j p_j}w;\beta^{q_j}), \qquad (2.30)$$

where $T_\epsilon^{(j)}(w) \equiv T^{(j)}(w;\epsilon)$ and $P_\epsilon(z) \equiv P(z;\epsilon)$. The roots of $T_\epsilon^{(j)}(w)$ are identical to those of $R_\beta^{(j)}(w)$, but the non-zero roots of the latter will have *regular expansions* in β of the form

$$w(\beta) = b_0 + b_1\beta + \cdots + b_N\beta^N + O(\beta^{N+1}). \qquad (2.31)$$

In summary:

> Every root of the polynomial (2.1) is of the form (2.2).
> The set of exponents (2.6) determines a set $\{p_1, \ldots, p_m\}$
> of proper values. With each proper value one introduces
> a new parameter β via (2.29) and an associated poly-
> nomial $R_\beta^{(j)}(w)$ defined by (2.30). The non-zero roots
> of $R_\beta^{(j)}(w)$ have regular perturbation expansions in β.
> The total number of non-zero roots of all the $R_\beta^{(j)}(w)$ is
> n. These yield expansions for each of the roots of (2.1).

To pull together all the ideas in the chapter, let us construct expansions for the roots of the polynomial

$$P_\epsilon(z) = 1 - \epsilon + \epsilon(2 + 3\epsilon^2)z - \epsilon^{-3}(16 - \epsilon)z^4$$
$$+ \epsilon^2(4 - \epsilon + \epsilon^3)z^6. \qquad (2.32)$$

In outline, proceed as follows:

1. Set $z = \epsilon^p w$ in (2.32) and determine the set of exponents:

$$P_\epsilon(\epsilon^p w) = 1 - \epsilon + \epsilon^{p+1}(2 + 3\epsilon^2)w - \epsilon^{-3+4p}(16 - \epsilon)w^4$$
$$+ \epsilon^{2+6p}(4 - \epsilon + \epsilon^3)w^6 \qquad (2.33)$$

$$E = \{0, 1 + p, -3 + 4p, 2 + 6p\}. \qquad (2.34)$$

2. Determine the pairs (p_j,e_j) of proper values and minimal exponents (with the aid of graphs or otherwise) and list the associated polynomials $T_\epsilon^{(j)}(w)$:

$$(3/4,0): T_\epsilon^{(1)}(w) = 1 - \epsilon + \epsilon^{7/4}(2 + 3\epsilon^2)w - (16 - \epsilon)w^4$$
$$+ \epsilon^{13/2}(4 - \epsilon + \epsilon^3)w^6 \qquad (2.35)$$

$$(-5/2, -13): T_\epsilon^{(2)}(w) = \epsilon^{13}(1 - \epsilon) + e^{23/2}(2 + 3\epsilon^2)w \qquad (2.36)$$
$$- (16 - \epsilon)w^4 + (4 - \epsilon + \epsilon^3)w^6.$$

3. For each j, determine q_j, set $\epsilon = \beta^{q_j}$, and list the associated polynomial $R_\beta^{(j)}(w)$:

$$\epsilon = \beta^4: R_\beta^{(1)}(w) = 1 - \beta^4 + \beta^7(2 + 3\beta^8)w -$$
$$(16 - \beta^4)w^4 + \beta^{26}(4 - \beta^4 + \beta^{12})w^6 \qquad (2.37)$$

$$\epsilon = \beta^2: R_\beta^{(2)}(w) = \beta^{26}(1 - \beta^2) + \beta^{23}(2 + 3\beta^4)w$$
$$- (16 - \beta^2)w^4 + (4 - \beta^2 + \beta^6)w^6. \qquad (2.38)$$

4. Each $R_\beta^{(j)}(w)$ has non-zero roots of the form (2.31). Substitute this expression into the identity $R_\beta^{(j)}(w(\beta)) \equiv 0$, collect and equate to zero coefficients of like powers of β, and solve, one-by-one, for the unknown coefficients $b_0, b_1, \cdots$.

From (2.37), the roots of $R_\beta^{(1)}(w)$ will have the form

$$w(\beta) = b_0 + b_4\beta^4 + O(\beta^7). \qquad (2.39)$$

However, from (2.38), the roots of $R_\beta^{(2)}(w)$ will have the form

$$w(\beta) = b_0 + b_2\beta^2 + O(\beta^4). \qquad (2.40)$$

The reason that (2.40) does not have an $O(\beta^6)$ error term is that $w(\beta)$ must be raised to the fourth power which will produce terms of the order shown in (2.40).

From (2.37)

$$1 - 16b_0^4 = 0 \qquad (2.41)_0$$
$$-1 - 64b_0^3b_4 + b_0^4 = 0. \qquad (2.41)_4$$

Hence

$$b_0 = (1/16)^{1/4} = \frac{1}{2}e^{(k-1)i\pi/2}, \ k = 1,2,3,4 \qquad (2.42)_0$$
$$b_4 = -(15/64)b_0. \qquad (2.42)_4$$

And from (2.38),

$$-16b_0^4 + 4b_0^6 = 0 \qquad (2.43)_0$$
$$-64b_0^3b_2 + b_0^4 + 24b_0^5b_2 - b_0^6 = 0. \qquad (2.43)_2$$

Hence

$$b_0 = (4)^{1/2} = 2(-1)^k, \ k = 5,6 \qquad (2.44)_0$$
$$b_2 = (3/32)b_0. \qquad (2.44)_2$$

5. Write down the roots $z_1(\epsilon), \ldots, z_6(\epsilon)$ of (2.34) via the change of

variable $z = \epsilon^p w$:

$$z_k(\epsilon) = 1/2e^{(k-1)i\pi/2}[1 - (15/64)\epsilon + O(\epsilon^{7/4})] , \quad k = 1,2,3,4 \tag{2.45}$$

$$z_k(\epsilon) = 2(-1)^k \epsilon^{-5/2}[1 + (3/32)\epsilon + O(\epsilon^2)] , \quad k = 5,6. \tag{2.46}$$

Exercise 2.5. Compute explicitly the $O(\epsilon^2)$ term in (2.46).

Exercise 2.6. Compute the first two non-vanishing terms in the expansions of each of the roots of

(a) $1 + \epsilon^{-1}z + \epsilon^{-1}z^2 + z^3$

(b) $1 - 2z + z^2 + \epsilon z^5$

(c) $1 + 2z + z^2 + \epsilon^4 z^4 + \epsilon^7 z^5.$

Exercise 2.7. The governing equations of the Morley-Koiter theory of circular cylindrical shells can be reduced to a single eighth order partial differential equation. Separation of variables yields a constant coefficient *ODE* whose associated polynomial is

$$\epsilon^4(z^2 - m^2)^2(z^2 - m^2 + 1)^2 + z^4,$$

where $m = 0, 1, 2, \cdots$ and ϵ^2 is proportional to the thickness to radius ratio of the shell. Determine one term expansions with remainder estimates (i.e., O-terms) for each of the roots, taking note of the special cases $m = 0,1$. [Hint: The polynomial is of the form $a^2 + b^2 = (a + ib)(a - ib)$. As the roots of a polynomial with real coefficients must occur in complex-conjugate pairs, the roots of $a - ib$ will be the conjugates of the roots of $a + ib$. Also, each factor is quadratic in z^2].

CHAPTER III: SINGULAR PERTURBATIONS IN ORDINARY DIFFERENTIAL EQUATIONS

The simplest type of ODE is linear and has constant coefficients: finding its general solution hinges on finding the roots of the associated polynomial. If the coefficients in the DE depend on a parameter ϵ then so do the roots of this polynomial. In general, a study of the behavior of these roots is *not* sufficient to infer the behavior of the solution of the DE because

(1) A non-homogeneous term in the DE can give rise to a term in the general solution whose behavior depends only, in part, on the roots of the associated polynomial.

(2) The imposition of initial conditions (IC's) or boundary conditions (BC's) may result in the appearance of ϵ in the constants multiplying the homogeneouos terms in the general solution.

(3) The domain of interest may depend on ϵ.

(4) Most importantly, the solution of the DE will be a function of both ϵ and the independent variable, say t, which immediately raises the question of whether certain approximate solution are uniformly accurate for all values of t in the domain of interest.

Fortunately, when ϵ is small, characterizing the ϵ-dependence of the solution of a DE with IC's or BC's simplifies. The rest of this book is devoted to this task. This chapter is intended to acquaint you with two simple problems that are typical of those we shall study later.

First, we consider the *initial value problem* (IVP)

$$L_\epsilon(y) \equiv \ddot{y} + 2\epsilon\dot{y} + y = 0 \,,\, |\epsilon| < 1 \,,\, 0 < t \,,\, \dot{y} = dy/dt$$

$$y(0) = 0 \,,\, \dot{y}(0) = 1.$$

(3.1)

If ϵ is small, we might expect the solution of (3.1) to be approximated accurately by the solution of the *reduced problem*

$$L_0(Y) = \ddot{Y} + Y = 0 \,,\, 0 < t$$

$$Y(0) = 0, \, \dot{Y}(0) = 1,$$

(3.2)

which is

$$Y(t) = \sin t.$$

(3.3)

Substituting (3.3) into the *DE* in (3.1), we see that, on the average (here meaning over one period, $t = 2\pi$), the *magnitude* term $2\epsilon\dot{Y}$ is uniformly small of order ϵ, compared to the magnitudes of each of the terms $\ddot{Y}$ and Y. However,

> Small disturbances acting for long times can have large effects as we see upon comparing (3.3) with the exact solution of (3.1),

$$y(t, \epsilon) = \frac{e^{-\epsilon t} \sin(\sqrt{1 - \epsilon^2}t)}{\sqrt{1 - \epsilon^2}}.$$

(3.4)

(See Fig. 3.1)

Exercise 3.1. Verify (3.4).

Because of the factor $e^{-\epsilon t}$ in (3.4), the amplitude of the exact solution either grows without bound ($\epsilon < 0$) or decays to zero ($\epsilon > 0$) as $t \to \infty$. Thus on the semi-infinite domain $0 \leq t$, $Y(t)$ is *not* an accurate approximation to $y(t,\epsilon)$ and we say that problem (3.1) exhibits *a singularity in the domain*. Such behavior has no counterpart in the algebraic problems we studied in Chapter II, as the polynomial $z^2 + 2\epsilon z + 1$ associated with (3.1) is regular.

On any *finite* domain, $0 \leq t \leq T$, we may apply the mean value theorem to (3.4). With $y_\epsilon \equiv \partial y / \partial \epsilon$, we have

$$y(t, \epsilon) = y(t, 0) + y_\epsilon(t, \epsilon_*)\epsilon$$

$$= \sin t + e^{-\epsilon_* t}\left[-\frac{t\,\sin(\lambda_* t)}{\lambda_*} - \frac{\epsilon_* t\,\cos(\lambda_* t)}{\lambda_*} + \frac{\epsilon_*\,\sin(\lambda_* T)}{\lambda_*^3} \right]\epsilon \quad (3.5)$$

$$= \sin t + O(\epsilon),\ 0 \leq t \leq T,$$

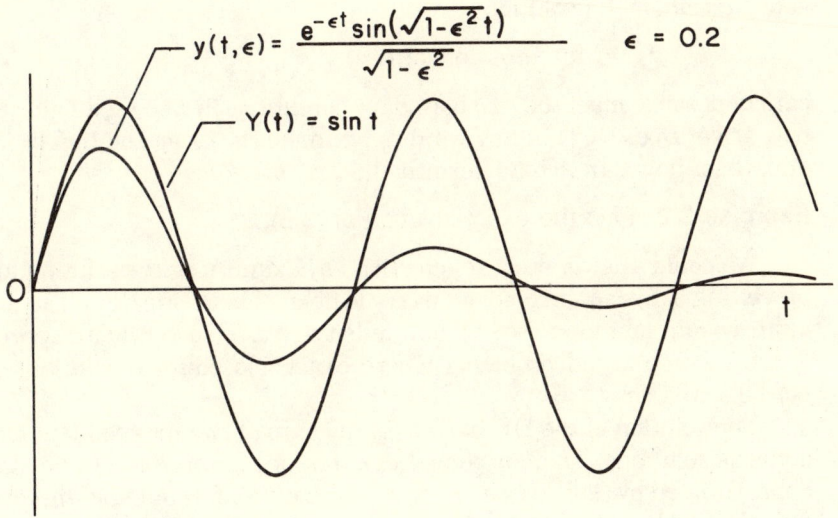

Fig. 3.1. Solution of Eq. (3.1), $y(t, \epsilon)$, and Eq. (3.2), $Y(t)$.

where $\lambda_* = \sqrt{1 - \epsilon_*^2}$ and ϵ_* is some number between 0 and ϵ. Thus we conclude that (3.1) is a *regular perturbation problem* if $0 \leq t \leq T$, but not if $0 \leq t$. How to tame singularities arising from (semi-)infinite domains will be studied in Chapter V.

A second interesting phenomenon occurs in the *boundary value problem* (BVP)

$$L_\epsilon(y) \equiv \epsilon \ddot{y} + \dot{y} + y = 0 , \, 0 < t < 1,$$

$$y(0) = 0 , \, y(1) = 1.$$

$$(3.6)$$

Immediately we see that we have a singular perturbation problem because the *reduced equation*

$$L_0(Y) = \dot{Y} + Y = 0 \tag{3.7}$$

is of *first order*. Its solution

$$Y(t) = Ae^{-t} , \, A \text{ constant,} \tag{3.8}$$

cannot possible meet the *two* boundary conditions in (3.6). Two questions now arise: Does (3.8) in any way approximate the exact solution of (3.6), and, if so, how can A be determined?

Exercise 3.2. Find the exact solution of (3.6).

We could answer our two questions by examining the solution to the preceding exercise. Instead, let us try to obtain, heuristically, an accurate *approximate* solution to see what it tells us. We do so to find an approach to more complicated problems where exact, closed-form solutions may not be available.

The solution of the DE in (3.6) is the sum of two independent homogeneous solutions. There is some hope that one of these might be closely approximated by (3.8) since (1), if we substitute (3.8) into the unreduced DE, the neglected term $\epsilon \ddot{Y}$ is *uniformly* small compared $\dot{Y}$ and Y (*i.e.*, compared to each of these terms individually, not compared to their sum, which is zero); and (2), the domain of the independent variable is finite. (We have deliberately considered a boundary value problem on a finite domain to avoid compounding the problems of a singular domain with that of a singular equation. In some physical problems both difficulties are present simultaneously).

Any second, independent solution of the DE in (3.6) must have the property that the first term $\epsilon \ddot{y}$, for some values of t, is of the same order of magnitude as one or more of the remaining terms; otherwise we would obtain again the reduced DE as a first approximation

The polynomial associated with the unreduced DE,

$$P_\epsilon(z) = \epsilon z^2 + z + 1, \tag{3.9}$$

is singular. To study its singular root, we make the substitution $z = \epsilon^{-1}w$ and obtain

$$\epsilon P_\epsilon(\epsilon^{-1}w) = w^2 + w + \epsilon. \tag{3.10}$$

Making the equivalent substitution $t = \epsilon\tau$ in (3.6), we have, with $y' = dy/d\tau$,

$$\epsilon L_\epsilon(y) = y'' + y' + \epsilon y = 0, \; 0<\tau<\epsilon^{-1},$$
$$y(0) = 0, \; y(\epsilon^{-1}) = 1. \tag{3.11}$$

If we now seek a solution of the transformed problem (3.11) such that the term ϵy is small compared to y'' and y', we are led to consider the simplified DE

$$W'' + W' = 0 \tag{3.12}$$

whose general solution is

$$W = B + Ce^{-\tau} = B + Ce^{-t/\epsilon}. \tag{3.13}$$

The constant B must be set to zero; otherwise ϵB will not be small compared to B'' and B' (which are zero), as was assumed.

Adding the two independent approximate solutions (3.8) and (3.13), we obtain

$$y \approx Ae^{-t} + Ce^{-t/\epsilon}, \tag{3.14}$$

which we hope will lead to a good approximation to the solution of (3.6). Imposition of the BC's in (3.6) yields

$$y \approx \frac{e}{1 - e^{1-1/\epsilon}} \, (e^{-t} - e^{-t/\epsilon})$$
$$\tag{3.15}$$
$$= e(e^{-t} - e^{-t/\epsilon}) + O(e^{-1/\epsilon}).$$

In Fig. 3.2, we compare the graph of the first term in the second line of (3.15) with the graph of ee^{-t}.

Exercise 3.3. Use the result of Exercise 3.2 to show that

$$|y(t,\epsilon) - e(e^{-t} - e^{-t/\epsilon})| = O(\epsilon).$$

Hint: Try to arrange the solution of Exercise 3.2 so that it looks as much like (3.15) as possible. Then drop transcendentally small terms.

The term $e^{-t/\epsilon}$ is called a *boundary layer* (or the boundary layer contribution to $y(t, \epsilon)$) because it is significant only in a narrow layer of width $O(\epsilon)$ near the boundary $t = 0$. As setting $\epsilon = 0$ in (3.6) often corresponds to replacing a physical model by a simpler one (*e.g.*, a shell by a membrane), we shall say that such BVP exhibits a *singularity in the model*. We can now answer, to some extent, the two questions we posed earlier. First, Ae^{-t} *does* approximate the solution of (3.6) outside a boundary layer at $t = 0$. For this reason, Ae^{-t} is called the *interior* contribution to $y(t, \epsilon)$. Second, because there is no boundary layer near $t = 1$, A *can* be determined from the BC at $t = 1$.

In the Chapters to follow, we shall develop systematic methods for reducing singular perturbation problems to regular ones. These will then be solved by constructing regular perturbation expansions.

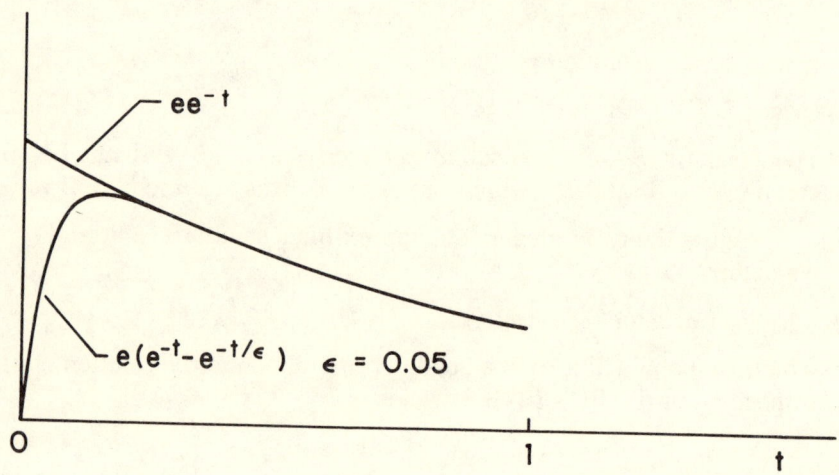

Fig. 3.2. Graphs of ee^{-t} and $e(e^{-t} - e^{-t/\epsilon})$.

Exercise 3.4. Explain how one or more of the points discussed at the beginning of the Chapter is illustrated by each of the following problems. Also explain why each is a singular perturbation problem. (Note that all may be solved exactly.)

(a) $\ddot{y} + (1 + \epsilon)y = 0 \, , \, 0 < t < \pi$

$y(0) = 0 \, , \, y(\pi) = \epsilon^p \, , \, p > 0.$

(b) $t^2\ddot{y} + t\dot{y} = 0 \, , \, \epsilon < t < 1$

$y(\epsilon) = 1 \, , \, y(1) = 0.$

(c) $\ddot{y} + y = \epsilon \cos t \, , \, 0 < t$

$y(0) = 0 \, , \, \dot{y}(0) = 1.$

CHAPTER IV: PERIODIC SOLUTIONS OF THE SIMPLEST NONLINEAR DIFFERENTIAL EQUATIONS. POINCARÉ'S METHOD

A Physical Model. Consider a frictionless cart of mass m attached to a spring that is anchored to a rigid wall (Fig. 4.1a). Let c denote the unstretched length of the spring and let x denote an additional displacement. From experiments on the spring we can construct a force-displacement curve. If the curve has a non-zero slope E at zero displacement, then it can be presented as a *dimensionless stress-strain curve*, as indicated in Fig. 4.1b.

Assume that the stress-strain relation is valid for any two springs having the same material properties and shape and let values of the stress F/E at a strain x/c be denoted by $f(x/c)$. Further, assume that tension in the spring produces extension and that compression produces contraction, *i.e.*, assume that $(x/c)f(x/c) > 0$, $x \neq 0$. The equation of motion of the cart then takes the form

$$m\frac{d^2x}{dT^2} = -F = -Ef(x/c). \tag{4.1}$$

To be specific we take as initial conditions

$$x = x_0, \frac{dx}{dT} = 0 \text{ at } T = 0. \tag{4.2}$$

We wish to study the motion of the cart, especially when the initial strain, x_0/c, is small. To this end it is convenient to nondimensionalize our IVP. A natural choice for a dimensionless displacement is x/x_0; another is the strain itself, x/c. The first choice makes the IC's parameter free; the second, the DE. Because $f(x/c)$ is essentially arbitrary at this stage, we shall take the strain as the dimensionless measure of displacement. Altogether then, if we set

$$x = cu, \; x_0 = c\epsilon, \; T = \sqrt{mc/E}, \tag{4.3}$$

our IVP takes the form

$$\ddot{u} + f(u) = 0, \quad 0 < t, \; u(0) = \epsilon, \; \dot{u}(0) = 0, \tag{4.4}$$

where $\dot{u} = du/dt$. Note that the IVP (1.87) for the large amplitude motion of a pendulum corresponds to taking $f(u) = \sin u$.

Before attempting to construct a suitable representation for the solution of (4.4), let us consider the simpler problem of determining, as a function of ϵ, the (dimensionless) period of oscillation of the motion. First,

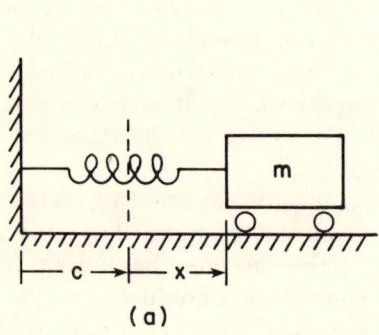

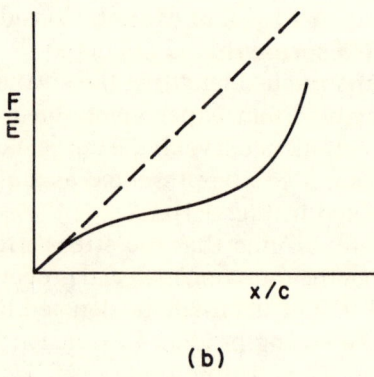

| (a) | (b) |

Fig. 4.1.(a) A mass attached to a nonlinear spring

Fig. 4.1.(b) Stress-strain relation for the spring.

though, we should assure ourselves that (4.4) *has* a periodic solution.

We begin by noting that if we multiple both sides of the DE in (4.4) by $\dot{u}$, the resulting equation can be written

$$\dot{K} + \dot{V} = 0, \tag{4.5}$$

where

$$K = \frac{1}{2}\dot{u}^2 \tag{4.6}$$

denotes the (dimensionless) *kinetic energy* of the cart and

$$V(u) = \int_0^u f(s)ds \tag{4.7}$$

denoted the (dimensionless) *potential energy* of the spring. Thus (4.5) implies the equation of conservation of energy,

$$K + V = C, \text{ a constant.} \tag{4.8}$$

From the IC's in (4.4), $C = V(\epsilon)$. Solving (4.8) for $\dot{u}$, we have

$$\dot{u} = \pm \sqrt{2[V(\epsilon) - V(u)]}\,. \tag{4.9}$$

To exploit our physical intuition, it is useful to recognize that (4.9) also represents the equation of motion of a bead that starts at rest from a height $V(\epsilon)$ and slides without friction along a wire of height $V(u)$. Here u measures distance *along* the wire, as indicated in Fig. 4.2.

As Fig. 4.2 suggests, a necessary condition for period motion is that there exist a constant $-\epsilon_*$ such that $V(-\epsilon_*) = V(\epsilon)$. For if the elevation of the wire to the left of $u = 0$ never reached $V(-\epsilon)$, the bead, once in motion, would continue to move forever to the left. (If the constant $-\epsilon_*$ exists, it must be unique. This follows from (4.7) and the fact that $uf(u)>0$, $u \neq 0$, which together imply that V is a strictly monotonically increasing function of $|u|$; *i.e.*, $V(u)$ is 1:1 on $u \geqslant 0$ and $u \leqslant 0$.)

To extract further information from (4.9), we introduce the very useful concept of a *phase plane*, the set of all ordered pairs of real numbers $(u,\dot{u})$. The *phase portrait* of the solution of (4.4) is the graph of the two functions defined by taking the + or − sign in (4.9). Since $V(\epsilon) - V(u)$ steadily decreases from $V(\epsilon)$ to 0 as u increases {decreases} from 0 to $\epsilon\{-\epsilon_*\}$, the graphs of these two functions form a single closed curve, as depicted in Fig. 4.3.

Equation (4.9) does not tell us how a point on the phase portrait varies with t. For this information, we return to the DE in (4.4) and write it in the form

$$(\dot{u}) = -f(u). \tag{4.10}$$

This equation and Fig. 4.1b imply that if u is positive, then $\dot{u}$ is a decreasing function of time, and *vice-versa*. Thus a point on the phase portrait moves clockwise, as indicated in Fig. 4.3.

The period of oscillation P is the time of one traverse. From (4.9) and Fig. 4.3,

$$P = \sqrt{2} \int_{-\epsilon_*}^{\epsilon} \frac{du}{\sqrt{V(\epsilon) - V(u)}}. \tag{4.11}$$

At the limits of integration the integrand has singularities. These must be integrable if the motion is to be periodic.

(To those who know some complex variable theory, we point out that if we regard u as complex, cut the u-plane from $-\epsilon_*$ to ϵ, assume that $V(u)$ can be extended analytically to a neighborhood of this cut, and choose the branch of the square root in (4.9) that reduces to $\sqrt{2V(\epsilon)}$ when

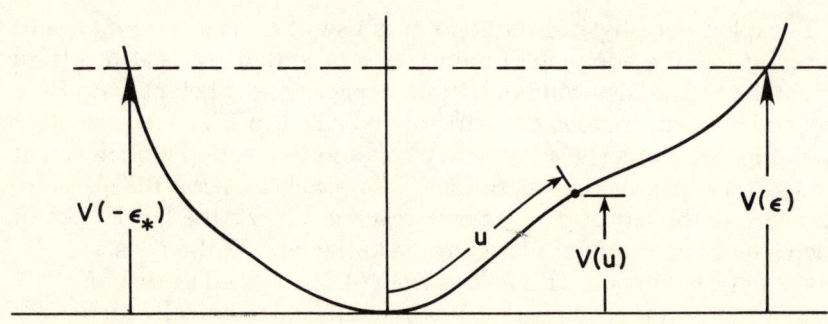

Fig. 4.2. Bead Sliding along a wire whose height is a function of distance along the wire.

$u = 0$, we obtain the elegant formula

$$P = \int_C \frac{du}{\sqrt{2[v(\epsilon) - v(u)]}},$$ (4.12)

where the integral is taken in a clockwise sense along a loop C surrounding the cut from $-\epsilon_*$ to ϵ, as shown in Fig. 4.4.)

Computation of $P(\epsilon)$ for $|\epsilon| \ll 1$ Directly by Formula. To be specific, we assume that

$$f(u) = u - \frac{1}{6}u^3 \, , \, u^2 < 6.$$ (4.13)[1]

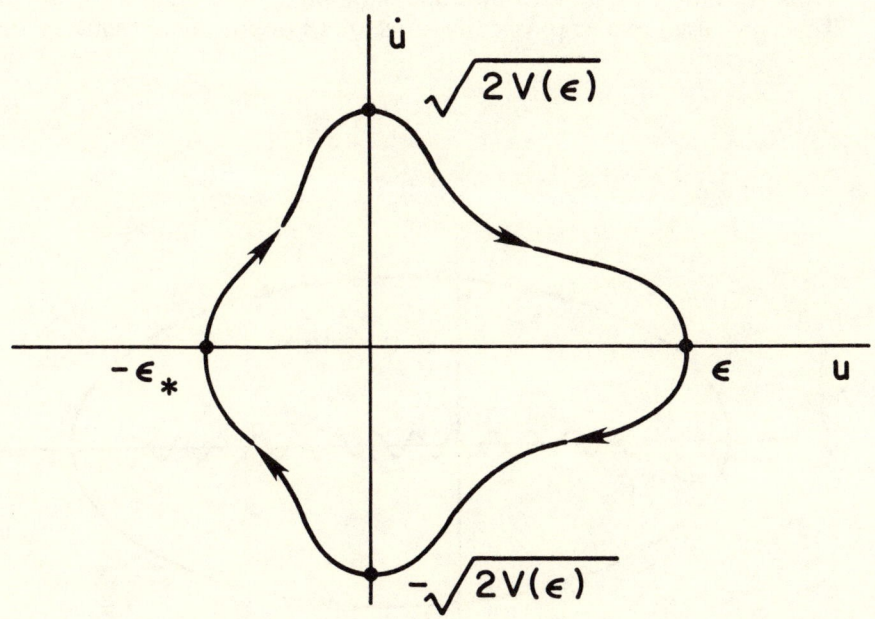

Fig. 4.3. Phase portrait associated with $K + V = V(\epsilon)$.

---

[1]$f(u)$ represents the first two terms in the Taylor Series expansion of $\sin u$ about $u = 0$.

From (4.7) the potential energy is

$$V(u) = \int_0^u (s - \frac{1}{6}s^3)ds = \frac{u^2}{2} - \frac{u^4}{24} . \tag{4.14}$$

In this case $\epsilon_* = \epsilon$, and (4.11) reduces to

$$P = 4\int_0^\epsilon \frac{du}{\{(\epsilon^2 - u^2)\,[1 - (1/12)(\epsilon^2 + u^2)]\}^{1/2}} . \tag{4.15}$$

To facilitate integration let

$$u = \epsilon \sin\alpha , \ 0 \leqslant \alpha \leqslant \pi/2 . \tag{4.16}$$

Then

$$P = 4\int_0^{\pi/2} \frac{d\alpha}{\{1 - (\epsilon^2/12)(1 + \sin^2\alpha)\}^{1/2}}$$
$$= 4 \int_0^{\pi/2}[1 + (\epsilon^2/24)(1 + \sin^2\alpha) + \cdots]d\alpha , \ \epsilon^2 < 6 \tag{4.17}$$
$$= 2\pi[1 + \epsilon^2/16 + O(\epsilon^4)] , \quad \epsilon^2 \leqslant 6 - \delta , \ \delta > 0.$$

Thus, the larger the initial amplitude ϵ, the longer the period of oscillation. This is what we expect physically, since (4.13) represents a "soft" spring,

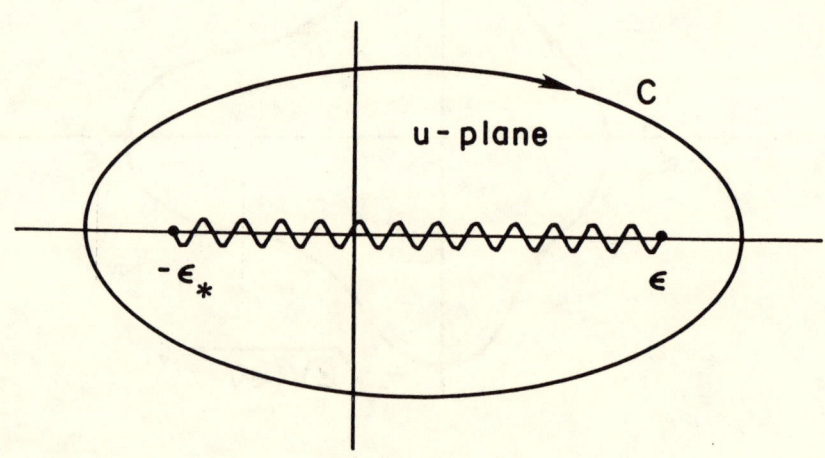

Fig. 4.4. Contour for the complex line integral in Eq. (4.12).

i.e., a spring whose stiffness decreases as it stretches.

Computation of $P(\epsilon)$ for $|\epsilon| \ll 1$ Indirectly by Poincaré's Method.
The solution (4.17) depended upon our being able to partially integrate
the DE in (4.4). Now we consider an alternate method that does not
require this. In addition to obtaining an approximate formula for $P(\epsilon)$,
we shall obtain a good approximation to the motion itself.

With (4.13) and the change of variables

$$u = \epsilon v, \epsilon^2 = \beta, \tag{4.18}$$

(4.4) reduces to

$$\ddot{v} + v - \frac{1}{6}\beta v^3 = 0, \; 0 < t, \; v(0) = 1, \; \dot{v}(0) = 0. \tag{4.19}$$

For a wide class of IVP's that includes (4.19), there is a theorem[2]
that says the solution is analytic in β for t and β sufficiently small. Thus
we are guaranteed the regular expansion

$$v(t,\beta) = v_0(t) + \beta v_1(t) + \cdots + \beta^N v_N(t) + R_{N+1}(t,\beta), \tag{4.20}$$

where, *for some* $T > 0$,

$$R_{N+1}(t,\beta) = O(\beta^{N+1}), \quad |t| < T. \tag{4.21}$$

*This useful result reduces the search for a function $v(t,\beta)$ of two variables
to a sequence of searches for functions $v_0(t), v_1(t), \cdots$ of one variable.* We
will not use the Theorem because its proof requires too many new tools,
and because it may fail to apply in more complicated problems where
physical experience suggests that a representation of the form (4.20) is
reasonable. Those who are unhappy with this attitude may regard all that
follows as merely a formal way of obtaining a conjectured form of solution
which then is to be justified rigorously.

Why the Unmodified Regular Expansion Won't Work. To pro-
ceed, we must assume that the derivatives $\dot{v}$ and $\ddot{v}$ also have expansions
with remainder estimates of the form (4.20) and (4.21). Substituting
these expansions along with (4.20) into (4.19), we have

$$(\ddot{v}_0 + \beta\ddot{v}_1 + \cdots + \beta^N\ddot{v}_N + \ddot{R}_{N+1}) + (v_0 + \beta v_1 + \cdots + \beta^N v_N$$
$$+ R_{N+1}) - (1/6)\beta(v_0 + \beta v_1 + \cdots + \beta^N v_N + R_{N+1})^3 = 0 \tag{4.22a}$$

$$v_0(0) + \beta v_1(0) + \cdots + \beta^N v_N(0) + R_{N+1}(0,\beta) = 1 \tag{4.22b}$$

$$\dot{v}_0(0) + \beta\dot{v}_1(0) + \cdots + \beta^N\dot{v}_N(0) + \dot{R}_{N+1}(0,\beta) = 0. \tag{4.22c}$$

---

[2]Coddington & Levinson, *Theory of Ordinary Differential Equations*, McGraw-Hill, 1955,
Theorem 8.4, pp. 36-37.

In (4.22a) we expand the cube in the second line and then collect coefficients of like powers of β to obtain

$$\ddot{v}_0 + v_0 + \beta[\ddot{v}_1 + v_1 - (1/6)v_0^3]$$

$$+ \beta^2[\ddot{v}_2 + v_2 - (1/2)v_0^2 v_1] + \cdots + O(\beta^{N+1}) = 0. \tag{4.23}$$

By the fundamental theorem of perturbation theory,, the coefficient of each power of β in (4.23) and the IC's (4.22b) and (4.22c) must vanish. This yields the sequence of IVP's

$$\ddot{v}_0 + v_0 = 0, \quad v_0(0) = 1, \dot{v}_0(0) = 0 \tag{4.24}_0$$

$$\ddot{v}_1 + v_1 = (1/6)v_0^3, \quad v_1(0) = \dot{v}_1(0) = 0 \tag{4.24}_1$$

$$\ddot{v}_2 + v_2 = (1/2)v_0^2 v_1, \quad v_2(0) = \dot{v}_2(0) = 0, \tag{4.24}_2$$

etc.

The solution of (4.24)$_0$ is

$$v_0(t) = \cos t. \tag{4.25}$$

Before proceeding farther, let us see how well $v_0(t)$ approximates $v(t,\beta)$. If the expansion (4.20) and remainder estimate (4.21) are valid, then for any *fixed* $T > 0$,

$$|v(t,\beta) - v_0(t)| = O(\beta), \, |t| < T. \tag{4.26}$$

The period of $v_0(t)$ is 2π whereas (4.17) shows that the period of $v(t,\beta)$ depends on β. Thus as t increases, $v_0(t)$ and $v(t,\beta)$ grow slowly out of phase (Fig. 4.5). Indeed, if $t = O(\beta^{-1})$ then, by (4.17), all we can assert is that $|v(t,\beta) - v_0(t)| = O(1)$. Clearly we are dealing with a perturbation problem in which there is a *singularity in the domain*. Will taking a two term approximation to $v(t,\beta)$ improve things?

Inserting (4.25) into the DE in (4.24)$_1$, we obtain

$$\ddot{v}_1 + v_1 = \frac{\cos t}{8} + \frac{\cos 3t}{24} \tag{4.27}$$

upon using the trigonometric identity

$$\cos^3 t = \frac{3}{4} \cos t + \frac{1}{4} \cos 3t. \tag{4.28}$$

The solution of (4.27) that meets the IC's in (4.24)$_1$ is easily found to be

$$v_1(t) = \frac{t \sin t}{16} + \frac{\cos t - \cos 3t}{192}. \tag{4.29}$$

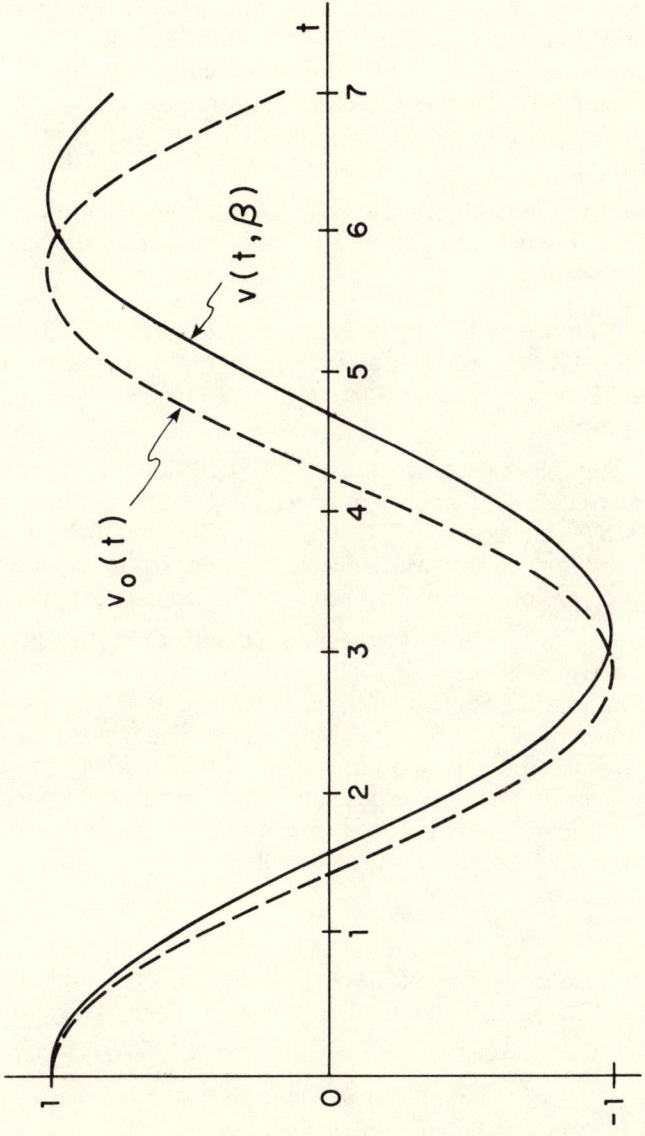

Fig. 4.5. Graphs of $v(t,\beta)$ and $v_0(t)$ showing how the two functions drift out of phase.

For any fixed $T > 0$, (4.21) tells us that

$$|v(t,\beta) - v_0(t) - \beta v_1(t)| = O(\beta^2). \tag{4.30}$$

However, it is clear from (4.29) that *on the full interval of interest, $t > 0$, $v_0(t) + \beta v_1(t)$ is a poorer approximation to $v(t,\beta)$ than $v_0(t)$ alone.* The culprit is $t \sin t$ which, unlike the exact solution, is non-periodic and grows without bound as $t \to \infty$. Moreover, the approximation $v_0(t) + \beta v_1(t)$ fails to reflect in any simple way the variation with β of the period of oscillation.

Exercise 4.1. Compute the period P by observing that P equals 4 times the value of t at which $v_0(t) + \beta v_1(t)$ first crosses the t-axis. Compare results with (4.17).

If we interpret (4.27) as the equation for the forced motion of a linear, undamped oscillator, then the factor $t \sin t$ in (4.29) represents the response of the oscillator to a resonant driving force, $(1/8)\cos t$.

In summary

> Though the solution of the IVP (4.19) admits a regular expansion, the remainder $R_{N+1}(\beta,t)$ is not uniformly $O(\beta^{N+1})$ for $0 < t$. To get an approximation to the exact solution, useful over times long compared to the period of oscillation, we must seek another type of expansion.

Poincaré's Method of Strained Coordinates. Let us consider our first approximation to $v(t,\beta)$. The function $v_0(t) = \cos t$ agrees with $v(t,\beta)$ in amplitude, but has a slightly shorter period. It is this discrepancy that drives $v_0(t)$ and $v(t,\beta)$ slowly apart. To resolve the difference suppose that we replace the first approximation by $\cos\lambda t$, where λ is a constant. Then *there must exist a value of* λ such that $\cos\lambda t$ and $v(t,\beta)$ have identical periods. Poincaré's method is a systematic procedure for determining successively more accurate approximations to λ.

The variable

$$T = \lambda t \tag{4.31}$$

is called a *strained variable*. The first step in Poincaré's method is to introduce T as the new independent variable. Then (4.19) takes the form

$$\lambda^2 v'' + v - \frac{1}{6}\beta v^3 = 0 , \quad 0 < T , v(0) = 1, v'(0) = 0, \tag{4.32}$$

where $v' = dv/dT$. Next, because $\lambda \to 1$ as $\beta \to 0$, it is natural to assume that λ has a regular expansion of the form

$$\lambda(\beta) = 1 + \beta\lambda_1 + \beta^2\lambda_2 + \cdots + \beta^N\lambda_N + O(\beta^{N+1}), \tag{4.33}$$

where the unknown constants $\lambda_1, \lambda_2, \cdots$ are to be found. Finally, we assume that the solution of (4.32) itself can be represented in the form

$$v(T,\beta) = z_0(T) + \beta z_1(T) + \beta^2 z_2(T) + \cdots + \beta^N z_N + R_{N+1}(T,\beta),$$
(4.34)

where

$$R_{N+1} = O(\beta^{N+1}),$$
(4.35)

uniformly in T. Substituting (4.33) and (4.34) into (4.32) and equating to zero the coefficients of successive powers of β, we obtain the following sequence of DE's and IC's.

$$\beta^0 : z_0'' + z_0 = 0 , \quad z_0(0) = 1, z_0'(0) = 0$$
(4.36)$_0$

$$\beta^1 : z_1'' + z = \frac{1}{6} z_0^3 - 2\lambda_1 z_0'' , z_1(0) = z_1'(0) = 0,$$
(4.36)$_1$

$$\beta^2 : z_2'' + z_2 = \frac{1}{2} z_0^2 z_1 - 2\lambda_1 z_1'' - (\lambda_1^2 + 2\lambda_2)z_0''$$
$$z_2(0) = z_2'(0) = 0, \quad (4.36)_2$$

etc.

The solution of (4.36)$_0$ is

$$z_0(T) = \cos T .$$
(4.37)

Substituting this into the DE in (4.36)$_1$ and using the trigonometric identity (4.28), we obtain

$$z_1'' + z_1 = (1/8 + 2\lambda_1)\cos T + (1/24)\cos 3T.$$
(4.38)

The $\cos T$-term on the right hand side of (4.38)—a *resonance-producing term*—will give rise to a term proportional to $T \sin T$ in the solution for z_1. Recall that it was such a term in our earlier solution for $v_1(t)$ that made $v_0(t) + \beta v_1(t)$ such a poor approximation to $v(t,\beta)$. However, now we have the constant λ_1 at our disposal, and we shall choose it to suppress the resonance-producing term in (4.38). That is, we take

$$\lambda_1 = -\frac{1}{16} .$$
(4.39)

With λ_1 in hand, our first approximation to $v(T,\beta)$ takes the form

$$z_0(T) = \cos T = \cos\lambda t = \cos[1 - (1/16)\beta + O(\beta^2)]t.$$
(4.40)

The period P of oscillation of z_0 [which we have *required* be equal to that of $v(t,\beta)$] satisfies the relation

$$[1 - (1/16)\beta + O(\beta^2)]P = 2\pi,$$
(4.41)

i.e.,

$$P = 2\pi[1 + (1/16)\beta + O(\beta^2)] , \tag{4.42}$$

which agrees exactly with (4.17).

Now comes an extremely important observation. If we wish to use the results we have obtained so far, namely $z_0(T)$ and λ_1, and *no more*, the best approximation we could use for $v(t,\beta)$ would be

$$v(t,\beta) \approx \cos[1 - (1/16)\beta]t . \tag{4.43}$$

Recall that the first approximation solution, $\cos t$, gave an error of $O(\beta)$ for $t = O(1)$, but a larger error of $O(1)$ for $t = O(\beta^{-1})$. However, (4.43) *now gives an error of only* $O(\beta)$ *for* $t = O(\beta^{-1})$ *and does not lead to an error of* $O(1)$ *until* $t = O(\beta^{-2})$.

It seems clear what to expect as we compute higher and higher approximations to $v(t,\beta)$. The right hand sides of each of the DE's in (4.36) will contain a resonance-producing term which must be suppressed by a proper choice of $\lambda_2, \lambda_3, \cdots$. In this way we obtain successively better approximations to the period P. At the same time, the knowledge of $z_2(T), z_3(T), \cdots$ allows the motion itself to be computed more and more accurately.

For example, if we compute $z_1(T)$ but not λ_2, we can expect $z_0(T) + \beta z_1(T)$ to approximate $v(t,\beta)$ to within an error of $O(\beta^{-2})$ for $t = O(\beta^{-1})$. If we also determine λ_2, we obtain the same degree of approximation to $v(t,\beta)$ but for $t = O(\beta^{-2})$, *etc.*

Exercise 4.2. Compute $z_1(T)$ and λ_2. Use your value of λ_2 to compute the next term in the expansion for the period given by (4.17). Check your result by direct computation from the second line of (4.17).

In Appendix B we *prove* that the remainder in (4.34) is uniformly small for all $T > 0$. That is, (4.34) is not an assumption but a consequence of (4.31) to (4.33).

In summary

> By partially accounting for the effect of the small parameter β in the first approximation to $v(t,\beta)$, *i.e.*, by replacing t by $T = \lambda(\beta)t$, we have reduced a singular perturbation problem to a regular one. However, there is a limitation. If only $\lambda_1,\ldots,\lambda_k$ have been computed, then (4.34) can be used to approximate $v(t,\beta)$ only if $t = O(\beta^{-k})$. Since $|\beta| \ll 1$, this may be no real limitation in practice.

Exercise 4.3. Consider the IVP

$$\ddot{u} + u + \epsilon u^2 = 0, \quad 0 < t, u(0) = 1, \dot{u}(0) = 0.$$

(a) For what range of values of ϵ do you expect periodic solutions to exist?

(b) If $|\epsilon| << 1$, try to solve for $u(t,\epsilon)$ by assuming a *regular* perturbation expansions in powers of ϵ. Carry your calculations far enough to indicate where and why the assumed expansion fails.

(c) Apply Poincaré's method to avoid the difficulties you found in (b).

Exercise 4.4. Use Poincaré's method to construct the first term in the regular expansion of the solution to the IVP

$$\ddot{v} + v + \frac{1}{2}\beta v |v| = 0, \quad |\beta| << 1, 0 < t, v(0) = 1, \dot{v}(0) = 0.$$

Hint: $\cos t |\cos t|$ is an even function of period 2π and hence can be represented by the Fourier series $\Sigma_0^\infty a_n \cos(nt)$, where

$$a_0 = (1/\pi) \int_0^\pi \cos t |\cos t| dt$$

$$a_n = (2/\pi) \int_0^\pi \cos(nt)\cos t |\cos t| dt, \quad n \geq 1.$$

Forced Motion of a Nonlinear Oscillator. Suppose that the frictionless spring-mass system of Fig. 4.1a is subject to an external, time-dependent force. Its dimensionless equation of motion can then be written as

$$\ddot{u} + f(u) = g(t), \quad 0 < t. \tag{4.44}$$

If $f(u)=u$, *i.e.*, if the DE is linear, then the solution of (4.44) is the sum of an homogeneous solution plus a particular solution:

$$u = u_h(t) + u_p(t)$$

$$= u(0)\cos t + \dot{u}(0)\sin t + \int_0^t \sin(t - \tau)g(\tau)d\tau. \tag{4.45}$$

The last term on the right of (4.45) is called the *forced response*.

Exercise 4.5. Verify that (4.45) satisfies (4.44) if $f(u) = u$ and that $u_p(0) = \dot{u}_p(0) = 0$.

If for a general f we set $f(u)=u+[f(u)-u]$ in (4.44) and take the term in brackets to the right side, then (4.45) becomes a nonlinear integral equation for u:

$$u = u(0)\cos t + \dot{u}(0)\sin t + \int_0^t \sin(t - \tau)[u(\tau) - f(u(\tau)) + g(\tau)]d\tau$$

(4.46)

$$\equiv \kappa(u).$$

Assuming $|u| \ll 1$, we might attempt to obtain a sequence of approximate solutions to (4.46) by the iteration process $u_{n+1} = \kappa(u_n)$, $u_0 = 0$. Alternatively, we might ask: can Poincaré's method be used to obtain approximate solutions? In general the answer is no, but in an important, special case that we shall now discuss, the answer is yes.

If the dimensionless, external force is harmonic, say

$$g(t) = \epsilon \sin \omega t,$$

(4.47)

then if $f(u) = u$, (4.45) yields

$$u_p = \epsilon \frac{\omega \sin t - \sin \omega t}{\omega^2 - 1}$$

$$\to \frac{1}{2}\epsilon(\sin t - t \cos t) \text{ as } \omega \to 1.$$

(4.48)

Thus if $\omega \neq 1$, the forced response of the linear spring-mass system is harmonic, but if $\omega = 1$, the forced response is non-harmonic with an amplitude that grows without bound as $t \to \infty$. Such resonance is of great practical concern in elastic structures.

Invariably, if an elastic structure undergoes large, non-rigid deformations, nonlinear internal forces appear. Thus in the simplest of all elastic models, the spring-mass system, it is natural to ask what happens to solutions of

$$\ddot{u} + f(u) = \epsilon \sin \omega t$$

(4.49)

if $\omega \approx 1$. To be concrete, let us assume that $f(u)$ is odd and has a representation for small u of the form

$$f(u) = u + ku^3 + O(u^5), \ k = O(1).$$

(4.50)[3]

If $\omega = 1$ and if at some instant $|u| \ll 1$, we expect the resonant term $\epsilon t \cos t$ to begin to dominate the motion. But as $|u|$ grows, the nonlinear term ku^3 in (4.50) will come into play, moving the natural frequency of the force-free system away from 1. Thus $|u(t;\omega, \epsilon)|$ need *not* approach ∞ as $\omega \to 1$.

For the nonlinear term ku^3 to counteract the resonance-producing effect of the forcing term $\epsilon \sin \omega t$, we must have $u = O(\epsilon^{1/3})$. This suggests the change of variable

$$u = (\beta/|k|)^{1/2}v, \ \beta \equiv |k|^{1/3}\epsilon^{2/3}.$$

(4.51)

---

[3]If $f(u) = u + ku^3$, (4.49) is called Duffing's equation.

Furthermore, as we wish to determine u not just at $\omega = 1$ but in a neighborhood of $\omega = 1$, we set

$$\omega^2 = 1 + \beta\sigma, \; T = \omega t. \tag{4.52}$$

The parameter σ may be called the *detuning*. Substituting (4.50) to (4.52) into (4.49) and dividing by $(\beta/|k|)^{1/2}$, we obtain

$$(1 + \beta\sigma)v'' + \text{sgn}(k)\beta v^3 = \beta\sin T + O(\beta^2), \quad v' = dv/dT. \tag{4.53}[4]$$

To obtain approximate solutions to (4.53) for arbitrary IC's consistent with the assumption that $v = O(1)$, we need the *two-scale method* discussed in the next chapter. However, if we are content with *any* solution of (4.53) that is of period 2π—take what IC's may come—then Poincaré's method suffices.

Thus we look for solutions of (4.53), uniformly valid in T, of the form

$$v(T;\beta) = v_0(T) + \beta v_1(T) + O(\beta^2), \tag{4.54}$$

the dependence on σ being understood. Substituting (4.54) into (4.53) and equating to zero coefficients of β^0 and β^1, we obtain

$$v_0'' + v_0 = 0 \tag{4.55}_0$$

$$v_1'' + v_1 = \sin T - \sigma v_0'' - \text{sgn}(k)v_0^3. \tag{4.55}_1$$

Substituting the solution of $(4.55)_0$,

$$v_0 = A_0\cos T + B_0\sin T, \tag{4.56}$$

into $(4.55)_1$ and using the trigonometric identity (4.28) plus

$$\cos^2 t\sin t = \frac{1}{4}\sin t + \frac{1}{4}\sin 3t \tag{4.57}$$

$$\cos t \sin^2 t = \frac{1}{4}\cos t - \frac{1}{4}\cos 3t \tag{4.58}$$

$$\sin^3 t = \frac{3}{4}\sin t - \frac{1}{4}\sin 3t, \tag{4.59}$$

we obtain

$$v_1'' + v_1 = -\frac{3}{4} A_0(A_0^2 + B_0^2)\cos T$$
$$+ [1 + \sigma B_0 - \frac{3}{4} B_0(A_0^2 + B_0^2)]\sin T$$
$$+ \frac{1}{4}A_0(3B_0^2 - A_0^2)\cos 3T - \frac{1}{4}B_0(3A_0^2 - B_0)\sin 3T. \tag{4.60}$$

To avoid resonant terms in v_1, the coefficients of $\cos T$ and $\sin T$ must be set to zero. This yields

---

[4]$\text{sgn}(x)$ is the signum or sign function: $\text{sgn}(x) = 1, x > 0, \text{sgn}(0) = 0, \text{sgn}(x) = -1$, $x < 0$.

$$A_0 = 0\,,\ 1 + \sigma B_0 - \frac{3}{4}B_0^{\,3} = 0.$$ (4.61)

Fig. 4.6 is a graph of B_0 *vs.* σ (computed by regarding σ as a function of B_0). Note that, depending on the value of $\sigma_c = (3/2)^{4/3} = 1.717\cdots$, there may be one, two, or three solutions to (4.49) of the form

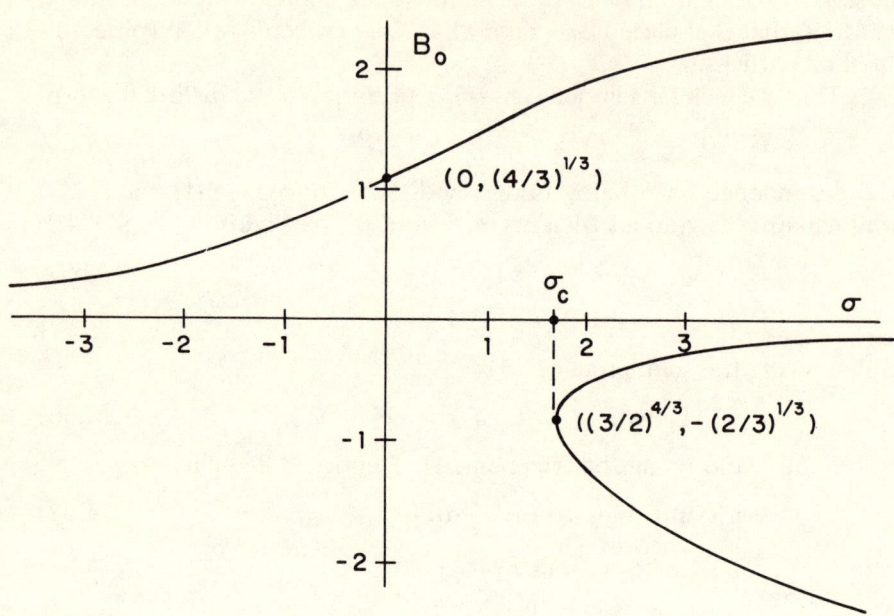

Fig. 4.6. Amplitude of the harmonic solutions of Duffing's equation as a function of the detuning.

$$u = (\beta/|k|)^{1/2}[B_0(\sigma)\sin \omega t + O(\beta)].\tag{4.62}$$

We have taken only the first step in the analysis of Duffing's equation. The treatment of arbitrary IC's, the inclusion of damping, and the elucidation of the behavior of the oscillator in a neighborhood of $\sigma = \sigma_c$ may be found in more detailed works on perturbation theory.[5]

Exercise 4.6. Suppose that $\omega \approx 3$ and that $k < 0$. Set $\omega = 3(1 + \beta\sigma)$, $T = (1 + \beta\sigma)t$ and show that there exist solutions of (4.49) of the form

$$u = (\beta/|k|)^{1/2}[2\sin(\frac{1}{3}\omega t) + O(\beta)].$$

These are called *sub-harmonic* solutions.

[5]See the two books by Nayfeh, *Introduction to Perturbation Methods*, Wiley, 1981 and *Perturbation Methods*, Wiley, 1973.

[3]If $f(u) = u + ku^3$, (4.49) is called Duffing's equation.

CHAPTER V:
INTRODUCTION TO THE TWO-SCALE METHOD

The Damped Linear Oscillator. Assume that the relative velocity of a pendulum swinging in air is small enough for the drag on the bob to be viscous and proportional to the velocity. If the only other damping is assumed to come from a torque at the pivot proportional to the angular velocity, then the equation of motion is

$$ml \frac{d^2\theta}{dt_*^2} = - mg \sin \theta - c \frac{d\theta}{dt_*} \ , \tag{5.1}$$

where m is the mass of the bob, l is the length of the pendulum, θ is its angular displacement from the vertical, g is the gravitational constant, c is a damping factor, and t_* is the time. Suppose that the pendulum, hanging at rest, is struck by a hammer at $t_* = 0$. Then the IC's are

$$\theta = 0 \ , \ d\theta/dt_* = \omega \ , \tag{5.2}$$

where $ml\omega$ is the impulse imparted by the hammer.

To work with the simplest model possible, assume that $|\theta(t_*)| \ll 1$, replace $\sin \theta$ by θ and nondimensionalize as follows:

$$t_* = t \sqrt{l/g} \ , \ 2\epsilon = c/(m \sqrt{lg}) \ , \ \theta = v\omega\sqrt{l/g}. \tag{5.3}$$

Then, with $(\ )^{\cdot} = d(\ )/dt$, the DE (5.1) and IC's (5.2) reduce to the IVP

$$\ddot{v} + 2\epsilon\dot{v} + v = 0 \ , \quad 0 < t \ , \ v(0) = 0 \ , \ \dot{v}(0) = 1. \tag{5.4}$$

Exercise 5.1. Show that the exact solution of (5.4) is

$$v(t,\epsilon) = \frac{e^{-\epsilon t}}{\sqrt{1-\epsilon^2}} \sin(t \sqrt{1-\epsilon^2}) \ . \tag{5.5}$$

Why an Unmodified Regular Expansion is No Good. Ignoring for the moment Exercise 5.1, let us see what follows if we assume that 1), the solution to (5.4) has the regular expansion

$$v(t,\epsilon) = v_0(t) + \epsilon v_1(t) + \cdots + \epsilon^N v_N(t) + R_{N+1}(t,\epsilon), \tag{5.6}$$

that 2), $\ddot{v}_0, \ddot{v}_1, \cdots \ddot{v}_N$, and $\ddot{R}_{N+1}$ exist, and that 3),

$$R_{N+1}, \dot{R}_{N+1}, \ddot{R}_{N+1} = O(\epsilon^{N+1}). \tag{5.7}$$

Substituting (5.6) into (5.4) and invoking The Fundamental Theorem, we get the following sequence of IVP's:

$$\epsilon^0: \ddot{v}_0 + v_0 = 0, \quad v_0(0) = 0, \dot{v}_0(0) = 1 \tag{5.8}_0$$

$$\epsilon^1: \ddot{v}_1 + v_1 + 2\dot{v}_0 = 0, \quad v_1(0) = \dot{v}_1(0) = 0, \tag{5.8}_1$$

etc. The solution of $(5.8)_0$ is

$$v_0(t) = \sin t. \tag{5.9}$$

Hence $(5.8)_1$ reduces to

$$\ddot{v}_1 + v_1 = -2\cos t, v_1(0) = \dot{v}_1(0) = 0, \tag{5.10}$$

whose solution is easily found to be

$$v_1(t) = -t \sin t. \tag{5.11}$$

So far, we have

$$v(t,\epsilon) = \sin t - \epsilon t \sin t + R_2(t,\epsilon). \tag{5.12}$$

Taking a peek at Exercise 5.1, we see that (5.12) is simply a Taylor expansion in ϵ of the exact solution.

Exercise 5.2. For t *fixed* and $|\epsilon| \leq 1 - \delta$, $\delta > 0$, show that $|R_2(t,\epsilon)| < K \epsilon^2$. Note, however, that the term $t \sin t$ (called a secular term) grows without bound as $t \to \infty$. Hence R_2 is not uniformly small in t.

Motivating the Two-Scale Method. To obtain a regular expansion that is *uniformly valid* we must, somehow, pull a part of the effect of ϵ into the first approximation to v. This is what the two-scale method of Cole and Kevorkian does. To provide physcial motivation, let us introduce the dimensionless energy

$$E = \frac{1}{2}(\dot{v}^2 + v^2). \tag{5.13}$$

Then with the aid of the DE in (5.4), we have

$$\dot{E} = -2\epsilon \dot{v}^2. \tag{5.14}$$

The IC's in (5.4) yield $E(0) = \frac{1}{2}$, so, upon integrating both sides of (5.14), we get

$$E(t) = \frac{1}{2} - 2\epsilon \int_0^t \dot{v}^2 ds. \tag{5.15}$$

Hence, E decreases monotonically.

Let us estimate the relative decay of E over one period. To within an error of order ϵ, we may replace $\dot{v}$ by $\dot{v}_0 = \cos t$ on the right side of (5.15). Thus

$$\frac{E(2\pi) - E(0)}{E(0)} \approx -4\epsilon\int_0^{2\pi}\cos^2 t\, dt = -4\epsilon\pi \tag{5.16}$$

or, roughly,

$$\frac{\Delta E}{E} \approx -2\epsilon\Delta t . \tag{5.17}$$

Integrating and setting $E(0) = \tfrac{1}{2}$, we get

$$E(t) \approx \frac{1}{2}\, e^{-2\epsilon t} . \tag{5.18}$$

Thus E decays to roughly $1/e$ of its initial value over a time $t = O(\epsilon^{-1})$.

This analysis shows that the solution for the damped linear oscillator contains *two* time scales, the period of oscillation, $t \approx 2\pi = O(1)$, and the energy "half life," $t \approx (2\epsilon)^{-1} = O(\epsilon^{-1})$. These observations suggest that it may be fruitful to think of the exact solution v as depending on two independent variables, or *two scales, t* and

$$\tau = \epsilon t . \tag{5.19}$$

In the terminology of Cole and Kevorkian, t is the "fast" variable and τ the "slow" variable.

From an Ordinary to a Partial Differential Equation. Assuming $v = v(t,\tau,\epsilon)$, we have by the chain rule and (5.19),

$$\dot{v} = \frac{dv}{dt} = \frac{\partial v}{\partial t} + \frac{\partial v}{\partial \tau}\frac{d\tau}{dt}$$

$$= \frac{\partial v}{\partial t} + \epsilon\,\frac{\partial v}{\partial \tau} \tag{5.20}$$

$$\equiv v_t + \epsilon v_\tau.$$

Likewise,

$$\ddot{v} = \frac{d}{dt}\,(\dot{v}) = (v_t + \epsilon v_\tau)_t + (v_t + \epsilon v_\tau)_\tau\,\frac{d\tau}{dt}$$

$$= v_{tt} + 2\epsilon v_{t\tau} + \epsilon^2 v_{\tau\tau} . \tag{5.21}$$

Inserting (5.20) and (5.21) into the original IVP, (5.4), we obtain

$$v_{tt} + v + 2\epsilon(v_{t\tau} + v_t) + \epsilon^2(v_{\tau\tau} + 2v_\tau) = 0 , \qquad 0 < t,\tau \tag{5.22}$$

$$v(0,0,\epsilon) = 0 \, , \, v_t(0,0,\epsilon) + \epsilon v_t(0,0,\epsilon) = 1.$$

(Things get worse before they get better!).

The Modified Regular Expansion. Now assume that v has the uniformly valid expansion

$$v(t,\tau,\epsilon) = \overset{0}{v}(t,\tau) + \epsilon\overset{1}{v}(t,\tau) + \cdots + \epsilon^N\overset{N}{v}(t,\tau) + R_{N+1}(t,\tau,\epsilon),$$

$$R_{N+1} = O(\epsilon^{N+1}). \tag{5.23}$$

This expression, inserted into (5.22), implies the following sequence of *partial* differential equations (PDE's) and IC's:

$$\epsilon^0: \overset{0}{v}_{tt} + \overset{0}{v} = 0 \, , \quad \overset{0}{v}(0,0) = 0, \, \overset{0}{v}_t(0,0) = 1 \tag{5.24$_0$}$$

$$\epsilon^1: \overset{1}{v}_{tt} + \overset{1}{v} + 2(\underline{\overset{0}{v}_{t\tau}} + \overset{0}{v}_t) = 0, \, \overset{1}{v}(0,0) = 0, \, \overset{1}{v}_t(0,0) + \overset{0}{v}_\tau(0,0) = 0, \tag{5.24$_1$}$$

etc.

Comparing the DE in (5.24)$_1$ to that in (5.8)$_1$ we see that the underlined term is new. It is this term which will give us the necessary freedom to suppress secular terms in $\overset{1}{v}$.

Exercise 5.3. Find (5.24)$_2$, the equation obtained by setting the coefficient of ϵ^2 to zero.

Solutions. The PDE in (5.24)$_0$ looks, formally, like the ODE in (5.8)$_0$. Its general solution is therefore

$$\overset{0}{v}(t,\tau) = A_0(\tau)\cos t + B_0(\tau)\sin t \, . \tag{5.25}$$

The IC's in (5.24)$_0$ yield

$$A_0(0) = 0 \, , \, B_0(0) = 1 \, . \tag{5.26}$$

Note that we get no information on $A_0(\tau)$ and $B_0(\tau)$ from the lowest order problem except their initial values. Inserting (5.25) into the PDE in (5.24)$_1$, we have

$$\overset{1}{v}_{tt} + \overset{1}{v} = 2[(A_0' + A_0)\sin t - (B_0' + B)\cos t] \, . \tag{5.27}$$

The Key Argument. The right side of (5.27) gives rise to particular solutions for $\overset{1}{v}$ of the form

$$C_1(\tau)t \sin t \quad \text{and} \quad D_1(\tau)t \cos t. \tag{5.28}$$

To suppress such secular terms, we must set the coefficients of $\sin t$ and $\cos t$ to zero in (5.27). Thus A_0 and B_0 must satisfy the ODE's

$$A_0' + A_0 = 0 \, , \, B_0' + B_0 = 0. \tag{5.29}$$

The solution of these equations, satisfying (5.26), are

$$A_0(\tau) = 0 \, , \, B_0(\tau) = e^{-\tau}. \tag{5.30}$$

We have now not only guaranteed that $\overset{1}{v}$, as a function of t, is bounded, but have also determined $\overset{0}{v}$ explicitly. Thus,

$$
\begin{aligned}
v &= \overset{0}{v} + O(\epsilon) \\
&= e^{-\tau}\sin t + O(\epsilon) \\
&= e^{-\epsilon t}\sin t + O(\epsilon) ,
\end{aligned}
\tag{5.31}
$$

for *all* $t > 0$.

Exercise 5.4. Compute $\overset{1}{v}(t,\tau)$ explicitly.

Exercise 5.5. Compare (5.31), with the $\overset{1}{v}$ term included, against the exact solution given in Exercise 5.1 and explain the significance of the term $\tau e^{-\tau}$ in $\overset{1}{v}$.

Exercise 5.6. Compute the energy $E(t,\epsilon)$ exactly using the results of Exercise 5.1. Compare your expression with the heuristically derived approximation (5.18).

Two Scales and a Strained Variable Combined. The exact solution given in Exercise 5.1 to the IVP (5.4) shows that the origin of the slow time $\tau = \epsilon t$ is the factor $e^{-\epsilon t}$. The exact solution also reveals something that we failed to anticipate: the fast time is not quite t but rather the *strained time*

$$T = t\sqrt{1 - \epsilon^2}. \tag{5.32}$$

Damping alters the period of oscillation. Failure to introduce a strained time results in the appearance of terms such as $\tau^n e^{-\tau}$ in $\overset{n}{v}$. (See Exercise 5.5). These do no real harm. Still, as max $(\tau^n e^{-\tau}) = n^n e^{-n} = e^n (\ln n - 1)$, max $|\overset{n}{v}(t,\tau)|$ does grow with n. Moreover, since the real part of $\tau^n e^{i\tau}$ is $\tau^n \cos \tau$, $\tau^n e^{-\tau}$ does have the formal appearance of a resonance. Thus, for aesthetic reasons if no others, it would be nice to suppress the $\tau^n e^{-\tau}$s. You are asked to do this in Exercise 5.7 which represents a good test of your mastery of Chapter IV and what we have covered so far in this Chapter.

Exercise 5.7 will illustrate another important principle in perturbation theory.

> There is no unique way to distribute the original independent variable t between the two new variables T and τ. The rule of thumb is to use any arbitrariness to simplify the expansions as much as possible—an admittedly subjective criterion.

Although the exact solution to the IVP (5.4) suggests (5.32) as the simplest fast scale, Exercise 5.7 will not automatically produce this result. Rather, if it is assumed that

$$T = (1 + \lambda_1\epsilon + \cdots)t , \tag{5.33}$$

then it will be found that there is no way to determine λ_1. The explanation is that the exact solution can also be expressed in terms of a different fast time, say,

$$T_* = t\sqrt{1 - \epsilon^2} - \lambda_1\epsilon t = t\sqrt{1 - \epsilon^2} - \lambda_1\tau, \tag{5.34}$$

in which case (5.5) takes the form

$$\sqrt{1 - \epsilon^2}\, v(T_*, t) = e^{-\epsilon t}\sin(T_* + \lambda_1\tau) \tag{5.35}$$

$$= e^{-\epsilon t}[\cos(\lambda_1\tau)\sin T_* + \sin(\lambda_1\tau)\cos T_*].$$

Thus an arbitrary value of λ_1 is to be expected in Exercise 5.7, but $\lambda_1 = 0$ will give the simplest expression for v_0.

Exercise 5.7. (a) In (5.4) set

$$\tau = \epsilon t , \quad T = \lambda(\epsilon)t \tag{5.36}$$

and asume that $v = v(T,\tau,\epsilon)$ to derive a new IVP analogous to (5.22).

(b) Assume regular expansions for $v(T,\tau,\epsilon)$ and $\lambda(\epsilon)$ and substitute these into the IVP obtained in (a) to get a sequence of IVP's. Carry the analysis far enough to suppress the $\tau e^{-\tau}$ term in $\overset{2}{v}$.

(c) Check your results against the exact solution for $\overset{1}{v}$ given in Exercise 5.1.

Nonlinear Damping. We now examine a problem for which there exists no exact solution—a cubically damped oscillator (This example is' Cole's):

$$\ddot{v} + \epsilon\dot{v}^3 + v = 0 , \quad 0 < t , \quad v(0) = 0, \, \dot{v}(0) = 1 . \tag{5.37}$$

An attempt to solve (5.37) by assuming a regular perturbation expansion

$$v(t, \epsilon) = v_0(t) + \epsilon v_1(t) + \cdots \tag{5.38}$$

can quickly be shown to lead to secular terms in $v_1(t)$.

Exercise 5.8. Show that a regular perturbation expansion of the solution of (5.37) leads to secular terms.

An examination of the (approximate) rate of decay of the energy again suggests that we assume

$$v = v(t, \tau, \epsilon) , \quad \tau = \epsilon t . \tag{5.39}$$

With the aid of (5.20) and (5.21), the IVP (5.37) takes the form

$$v_{tt} + v + \epsilon(v_t{}^3 + 2v_{t\tau}) + \epsilon^2(3v_t^2 v_\tau + v_{\tau\tau}) + 3\epsilon^3 v_t v_\tau^2 + \epsilon^4 v_\tau^3 = 0 \tag{5.40}$$

$$v(0,0,\epsilon) = 0 , \quad v_t(0,0,\epsilon) + \epsilon v_\tau(0,0,\epsilon) = 1.$$

Exercise 5.9. Verify (5.40).

As with linear damping assume that

$$v(t,\tau,\epsilon) = \overset{0}{v}(t,\tau) + \epsilon \overset{1}{v}(t,\tau) + \cdots . \tag{5.41}$$

This series, substituted into (5.40), implies that

$$\epsilon^0 : \overset{0}{v}_{tt} + \overset{0}{v} = 0 , \quad \overset{0}{v}(0,0) = 0 , \overset{0}{v}_t(0,0) = 1. \tag{5.42}_0$$

$$\epsilon^1 : \overset{1}{v}_{tt} + \overset{1}{v} + \overset{0}{v}_t{}^3 + 2\overset{0}{v}_{t\tau} = 0 , \quad \overset{1}{v}(0,0) = 0 , \overset{1}{v}_t(0,0) + \overset{0}{v}_\tau(0,0) = 0, \tag{5.42}_1$$

etc.

Exercise 5.10. Find $(5.42)_2$.

The solution of $(5.42)_0$ is

$$\overset{0}{v}(t,\tau) = A_0(\tau)\cos t + B_0(\tau)\sin t , \tag{5.43}$$

where

$$A_0(0) = 0 , \; B_0(0) = 1 . \tag{5.44}$$

Substituting (5.43) into the PDE in $(5.42)_1$, we have

$$\overset{1}{v}_{tt} + \overset{1}{v} = 2[A_0'(\tau)\sin t - B_0'(\tau)\cos t] + [A_0(\tau)\sin t - B_0(\tau)\cos t]^3. \tag{5.45}$$

In the hope of avoiding much unnecessary algebra, we conjecture that $A_0(\tau) \equiv 0$. Then with the aid of the trigonometric identity (4.28) we may cast (5.45) into the form

$$\overset{1}{v}_{tt} + \overset{1}{v} = -(2B_0' + 3B_0^3/4)\cos t - (B_0^3/4)\cos 3t. \qquad (5.46)$$

Clearly, to avoid a resonant term in $\overset{1}{v}$, we must set

$$2B_0' + 3B_0^3/4 = 0. \qquad (5.47)$$

The solution of this nonlinear ODE, satisfying the IC $B_0(0) = 1$, is found easily to be

$$B_0(\tau) = \frac{1}{\sqrt{(3/4)\tau + 1}}. \qquad (5.48)$$

Hence, from (5.43),

$$\overset{0}{v}(t,\tau) = \frac{\sin t}{\sqrt{(3/4)\tau + 1}} = \frac{\sin t}{\sqrt{(3/4)\epsilon t + 1}}. \qquad (5.49)$$

Note that

(1) Damping is algebraic rather than exponential.

(2) The influence of ϵ has been brought into the first approximation solution $\overset{0}{v}(t,\tau)$ through the appearance of the slow time $\tau = \epsilon t$.

(3) As with linear damping, it was necessary to consider the IVP for $\overset{1}{v}$ before $\overset{0}{v}$ could be pinned down completely. With $B_0(\tau)$ in hand, we can now solve (5.46):

$$\overset{1}{v}(t,\tau) = A_1(\tau)\cos t + B_1(\tau)\sin t + \frac{\cos 3t}{32[(3/4)\tau + 1]^{3/2}}. \qquad (5.50)$$

Exercise 5.11. Carry the analysis far enough to determine $\overset{1}{v}$ explicitly. Are there any indications that a fast time $T = \lambda(\epsilon)t$ should have been introduced?

Exercise 5.12. Consider the following IVP for a quadratically damped oscillator:

$$\ddot{v} + \epsilon|\dot{v}|\dot{v} + v = 0 \ , \ 0 < t \ , \ v(0) = 1 \ , \ v(0) = 0.$$

Assuming that ϵ is small and positive, look for a solution of the form

$$v(t,\tau,\epsilon) = \overset{0}{v}(t,\tau) + \epsilon\overset{1}{v}(t,\tau) + \cdots,$$

where τ, the slow time, is to be suitably related to t. Carry your work far enough to determine $\overset{0}{v}$ explicitly. Hint: See Exercise 4.4.

Exercise 5.13. (This is an open problem, *i.e.*, we don't know the complete answer.) If ϵ is small and positive, construct a uniformly valid first approximation to the linear oscillator with damping proportional to time:

$$\ddot{v} + \epsilon t \dot{v} + v = 0 \,, 0 < t \,, v(0) = 0 \,, \dot{v}(0) = 1.$$

For a start, try the naive expansion

$$v(t,\epsilon) = v_0(t) + \epsilon v_1(t) + \cdots$$

and see what happens.

Exercise 5.14. Consider (1.90), the dimensionless equation for a beam under tension on an elastic foundation. Set $f(x) = 0$ and assume that the beam is semi-infinite and subject to the BC's

$$y(0) = 1 \,, y'(0) = 0 \,, y(x) \,, y'(x) \to 0 \text{ as } x \to \infty.$$

(a) Solve the BVP exactly for $0 < \epsilon < 1/2$. If $0 < \epsilon << 1$, the solution exhibits a very short scale and a very long one. What are these?

(b) Introduce a short length X and a long length ξ to account for the two disparate length scales found in (a) and assume that $y = y(X,\xi,\epsilon)$. Use the two-scale method to determine a uniformly valid first-approximation $\bar{y}(X,\xi)$ to $y(x,\epsilon)$. Compare your result with the exact solution.

CHAPTER VI:
THE WKB(J)[1]
APPROXIMATION

A tightly stretched string of length L under a tension T is shown in Fig. 6.1. The string has a mass/length $\rho(\xi)$, where ξ is distance along the string from the left end. If a particle initially at ξ undergoes a small transverse displacement $w(\xi,t)$, where t is time, then its equation of motion is

$$T\frac{\partial^2 w}{\partial \xi^2} = \rho\frac{\partial^2 w}{\partial t^2} \ . \tag{6.1}$$

A derivation of (6.1) may be found in a number of texts.

Exercise 6.1. Reproduce such a derivation.

Standing waves are, by definition, motions of the string such that each particle moves harmonically in time, *i.e.*,

$$w(\xi,t) = W\ (\xi)\cos\Omega t \ , \tag{6.2}$$

where Ω is the *frequency of vibration*. Substituting (6.2) into (6.1), we obtain the ODE

$$T\frac{d^2 W}{d\xi^2} + \Omega^2\rho W = 0 \ . \tag{6.3}$$

We shall require that the ends of the string remain fixed. Thus the solutions of (6.3) are subject to tbe BC's

$$W(0) = W(L) = 0 \ . \tag{6.4}$$

Nondimensionalization. Let

$$\xi = Lx \ , \ W = Ly \ , \ \rho = \bar{\rho}r \ , \ \omega^2 = \bar{\rho}\Omega^2 L^2/T, \tag{6.5}$$

---

[1] The initials stand for Wentzel, Kramers, Brillouin, and Jefferys who invented the method in the 1920's to deal with quantum mechanical problems. According to Steele ("Application of the WKB Method in Solid Mechanics," *Mechanics Today*, Vol. 3. Ed. S. Nemat-Nasser, Pergamon Press, 1976), priority should go to Carlini (1817), Green (1837) and Liouville (1837) but the name WKB is now traditional.

where $\bar{\rho} = \max \rho(\xi)$, $0 \leqslant \xi \leqslant L$. Then the DE (6.3) and BC's (6.4) reduce to

$$y'' + \omega^2 r(x)y = 0 \, , \, 0 < x < 1 \, , \, y(0) = y(1) = 0. \qquad (6.6)$$

The Eigenvalue Problem. The BVP (6.6) has the *trivial solution* $y(x) \equiv 0$. But if $r(x) > 0$ on $(0,1)$, then there exist an infinite number of discrete values of ω, $0 < \omega_0 < \omega_1 < \ldots$, such that for each non-negative integer n, (6.6) has a non-trivial solution $y_n(x)$. ω_n is called the nth *natural frequency of vibration* and $y_n(x)$, the *associated mode*[2]. Finding the pairs $\{\omega_n, y_n\}$ is called an *eigenvalue problem* (EVP). Values of $\lambda \equiv \omega^2$ for which (6.6) has non-trivial solutions are called *eigenvalues* and the associated solutions $y_n(x)$, *eigenfunctions*. (Because EVP's are homogeneous, eigenfunctions are determined only to within an arbitrary multiplicative factor.)

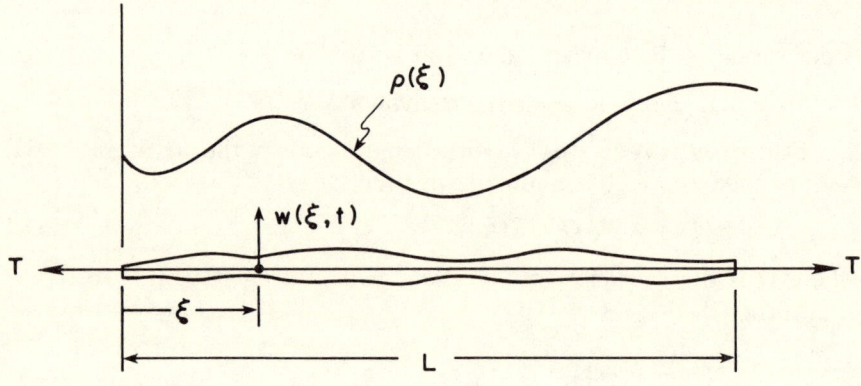

Fig. 6.1. A tightly stretched string of variable density.

[2]See, for example, Burkill, *The Theory of Ordinary Differential Equations*, 2nd Ed., Oliver and Boyd, 1962, Chapter III.

It may be shown, as indicated in Fig. 6.2, that $y_n(x)$ crosses the interval $(0,1)$ on the x-axis exactly n times, *i.e.*, the *nth mode has n nodes.*

The WKB Approximation for High Frequencies. Closed form solutions of (6.6) exist only for the simplest functions $r(x)$. However, if $\omega_n >> 1$, then between any two successive zeros of $y_n(x)$, $r(x)$ is nearly constant (see Fig. 6.2). This observation may be exploited as follows.

If r *were* a constant, then (6.6) would have solutions of the form

$$y_{\pm}(x,\omega) = Ae^{\pm i\omega x r^{1/2}} . \tag{6.7}$$

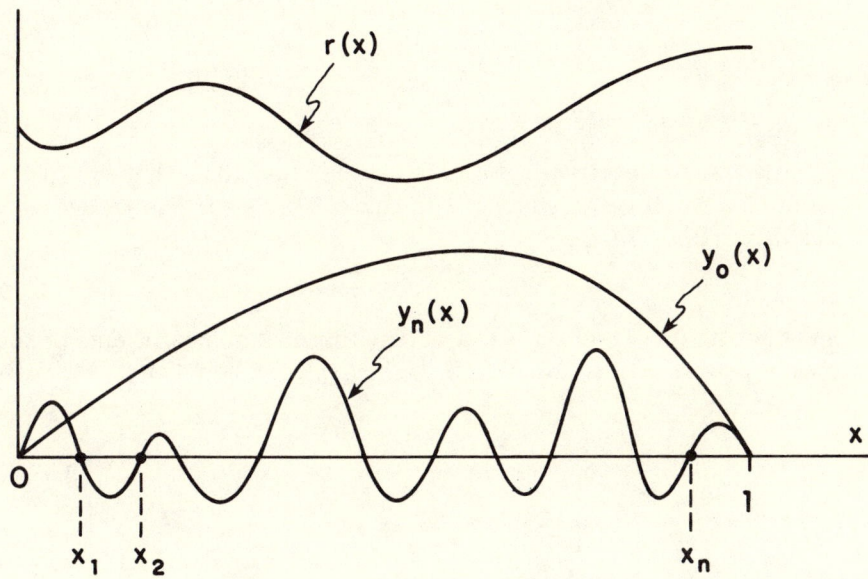

Fig. 6.2. Qualitative behavior of the eigenfunctions of Eq. (6.6).

This suggests that we assume

$$y(x,\omega) = e^{\omega g(x,\omega)} . \tag{6.8}$$

Then

$$y' = \omega g' e^{\omega g} \tag{6.9}$$

$$y'' = (\omega^2 g'^2 + \omega g'')e^{\omega g}, \tag{6.10}$$

and the DE in (6.6), upon division by ω^2, reduces to

$$\omega^{-1} g'' + g'^2 + r = 0. \tag{6.11}$$

As (6.11) is first order in g', we set

$$g = \int h dx , \tag{6.12}$$

so obtaining

$$\omega^{-1} h' + h^2 + r = 0. \tag{6.13}$$

Always, we start with the naive assumption that when a problem contains a small parameter (ω^{-1} in this case), its solution is regular in this parameter. Thus, we set

$$h(x,\omega) = h_0(x) + \omega^{-1} h_1(x) + \cdots . \tag{6.14}$$

Substituting (6.14) into (6.13) and equating to zero coefficients of successive powers of ω^{-1}, we obtain the sequence of equations

$$h_0^2 + r = 0 \tag{6.15$_0$}$$

$$2h_0 h_1 + h_0' = 0 , \tag{6.15$_1$}$$

etc. Since $r(x) > 0$, (6.15)$_0$ has the solution

$$h_0(x) = \pm i r^{1/2}(x) , \tag{6.16}$$

and (6.15)$_1$ yields

$$h_1(x) = -(1/4)[\ln r(x)]' . \tag{6.17}$$

So far,

$$y(x,\omega) = \exp[\omega \int h_0 dx + \int h_1 dx + O(\omega^{-1})]$$
$$= r^{-1/4}(x)\exp\omega[\pm i \int_a^x r^{1/2}(t)dt + O(\omega^{-2})] , \tag{6.18}$$

where the lower limit of integration a is to be chosen for convenience and multiplicative constants of integration have been set to 1. In terms of real quantities, the general homogeneous solution is of the form

$$y(x,\omega) = [r^{-1/4}(x) + O(\omega^{-1})]\{A\cos\omega[\int_a^x r^{1/2}(t)dt + O(\omega^{-2})]$$
$$+ B\sin\omega[\int_a^x r^{1/2}(t)dt + O(\omega^{-2})]\} \ , \tag{6.19}$$

where A and B are arbitrary real constants. It may be proved rigorously that (6.19) is an asymptotic expansion. (See Erdélyi, *Asymptotic Expansion*, Dover, 1953 p. 79.) When the O-terms are omitted, (6.19) is called the *WKB approximation*.

Exercise 6.2. Show that

$$h_2(x) = \pm \frac{i}{8r^{1/2}} \left[\frac{5}{4}\left(\frac{r'}{r}\right)^2 - \frac{r''}{r}\right]. \tag{6.20}$$

Exercise 6.3. Since $\exp[O(\omega^{-n})] = 1 + O(\omega^{-n})$, the expansions (6.18) can be written, alternatively, in the form

$$y(x,\omega) = A(x,\omega)\exp[\omega g(x)]$$
$$= [A_0(x) + \omega^{-1}A_1(x) + \ldots]\exp[\pm i\omega \int_0^x r^{1/2}(t)dt].$$

Derive the DE's satisfied by A_0 and A_1 and note the relation of A_1 to h_2.

First Order Solution to the Eigenvalue Problem. Setting $a = 0$ and substituting (6.19) into the BC's in (6.6), we find, in the limit as $\omega \to \infty$, that

$$A = 0 \ , \ B\sin[\omega\int_0^1 r^{1/2}(t)dt] = 0. \tag{6.21}$$

To avoid the trivial solution, the coefficient of B must vanish. This yields the asymptotic formula

$$\omega_n \sim \hat{\omega}_n \equiv \frac{(n+1)\pi}{\int_0^1 r^{1/2}(t)dt}, \ n \to \infty . \tag{6.22}$$

Hence,

$$y_n(x,\omega_n) \sim B_n r^{-1/4}(x)\sin[\hat{\omega}_n \int_0^x r^{1/2}(t)dt] \ , \tag{6.23}$$

where the B_n's are arbitrary constants.

Comparison with some Exact Solutions. If $r(x) = x^m$, $m > -2$, the exact solution of (6.6) may be expressed in terms of Bessel functions:

$$y_n(x,\omega_n) = B_n x^{1/2} J_p(\mu_n^{(p)} x^{1/2p}) \ , \tag{6.24}$$

where

$$p = \frac{1}{m+2} \ , \tag{6.25}$$

$\mu_n^{(P)}$ is the nth zero of $J_p(z)$, and

$$\omega_n = \mu_n^{(p)}/2p \ . \tag{6.26}$$

(See, for example, Hildebrand, *Advanced Calculus for Applications*, 2nd Ed., Prentice-Hall, 1976, p. 153.) From (6.22),

$$\hat{\omega}_n = (\tfrac{1}{2})(m + 2)(n+1)\pi \ . \tag{6.27}$$

If $m = 0$, (6.23) and (6.27) are exact, whereas if $m = \pm 1$, $\mu_n^{(p)}$ may be found in published tables. (*E.g.*, Watson, *Theory of Bessel Functions*, 2nd Ed., Cambridge University Press, 1962, Table VII). Table 6.1 compares values of ω_n from (6.26) with those given by (6.27).

Table 6.1 Exact and approximate natural frequencies of (6.6) with $r = x^{\pm 1}$.

	$m = -1$		$m = 1$	
n	$\omega_n = (1/2)\mu_n^{(1)}$	$\hat{\omega}_n = (1/2)(n+1)\pi$	$\omega_n = (3/2)\mu_n^{(1/3)}$	$\hat{\omega}_n = (3/2)(n+1)\pi$
0	1.9159	1.5708	4.3539	4.7124
1	3.5078	3.1416	9.0491	9.4248
2	5.0867	4.7124	13.756	14.137
3	6.6619	6.2832	18.465	18.850
4	8.2353	7.8540	23.176	23.562

Exercise 6.4. Show, formally, that with the change of dependent variable $w = vy$ and a proper choice of the function v, the DE

$$a(x)w'' + b(x)w' + [c(x) + \omega^2 d(x)]w = 0 \ , \ 0 < x < 1 \tag{6.28}$$

reduces to the so-called *normal form*

$$y'' + [q(x) + \omega^2 r(x)]y = 0 \ , \ 0 < x < 1. \tag{6.29}$$

Exercise 6.5. Use the results of Exercise 6.4 to reduce the following DE's to normal form.

(a) $y'' + 2\epsilon y' + \omega^2 y = 0$

(b) $(1 - x^2)y'' - 2xy' + \omega^2 y = 0$

(c) $y'' + xy' + (x^2 + \omega^2)y = 0.$

Exercise 6.6. If $r(x) > 0$ on $(0,1)$, show, formally, that the general solution of (6.29) is of the same form as (6.18).

Exercise 6.7. Consider the natural frequencies of a tightly stretched string composed of two strings of equal length, welded end-to-end, where the mass/length of one of the pieces is four times that of the other. It follows that $r(x) = 1/4$ on the interval $(0,1/2)$ and that $r(x) = 1$ on the interval $(1/2,1)$.

  (a) Compute the (dimensionless) natural frequencies by solving the DE in (6.6) on the two domains $(0,\frac{1}{2})$ and $(\frac{1}{2},1)$, subject to the BC's $y(0) = y(1) = 0$ and the junction conditions that y and y' be continuous at $x = \frac{1}{2}$.
  (b) Compute $\hat{\omega}_n$ from (6.22) and compare your answers with those obtained in part (a). To use a phrase of our youth, *the WKB approximation is not "phased" by discontinuities in the differential equation!* (This is a drawback as well, for if we study traveling waves, the WKB approximation, unless modified, does not predict the reflection that must occur at a discontinuity.)

Exercise 6.8. Find a change of independent variable $\xi = \xi(x)$, in terms of integrals of $a(x)/b(x)$, such that (6.28) reduces to the alternative normal form

$$\ddot{w} + [Q(\xi) + \omega^2 R(\xi)]w = 0 \ , \ \dot{w} = dw/d\xi \ . \tag{6.30}$$

Exercise 6.9. Reduce the DE's in Exercise 6.5 to the form (6.30). Compare the advantages (if any) of this form to (6.29).

Exercise 6.10.

  (a) Show that (6.6), (6.19), and (6.22) imply that

$$\omega_n = \hat{\omega}_n + O(n^{-1}). \tag{6.31}$$

  (b) Use the results of Exercise 6.2 to compute an $O(n^{-1})$ correction to $\hat{\omega}_n$.
  (c) Explain what goes wrong when you attempt to use the results obtained in (b) to compute an "improved" approximation to ω_n for $r(x) = x$ or $r(x) = \sin x$. (See Exercise 7.5).

The WKB Approximation for Non-Oscillatory Solutions. With the change of parameter

$$\omega = -i\epsilon^{-1}, \tag{6.32}$$

the DE in (6.6), upon multiplication by ϵ^2, takes the form

$$\epsilon^2 y'' - r(x)y = 0 \ , \ 0 < x < 1 \ , \ r(x) > 0 \ . \tag{6.33}$$

Exercise 6.11. Show that any solution of this DE can have at most one zero for $0 \leq x \leq 1$. Hint: Recall Rolle's Theorem (a special case of the mean value theorem) which states that if $y(x_1) = y(x_2), x_1 < x_2$, and if $y(x)$ is differentiable on (x_1, x_2), then there exists an x_* in (x_1, x_2) such that $y'(x_*) = 0$. Write $\epsilon^2 y'' = ry$ and integrate between appropriate limits to get a contradiction.

From (6.18) and (6.32) follows immediately the WKB approximation to the two solutions of (6.33):

$$y_{\pm}(x,\epsilon) \sim r^{-1/4}(x)\exp[\pm \epsilon^{-1}\int_0^x r^{1/2}(t)dt] \ . \qquad (6.34)\pm$$

If $0 < \epsilon << 1$, then ϵ^{-1} is large and the exponential factor in (6.34) associated with y_- decays rapidly from 1 towards 0 as x increases from 0. See Fig. 6.3. In the WKB approximation to y_+ it is convenient to have an exponential term that displays the same behavior with respect to the right end of the domain $(0,1)$. As an arbitrary constant times a homo-

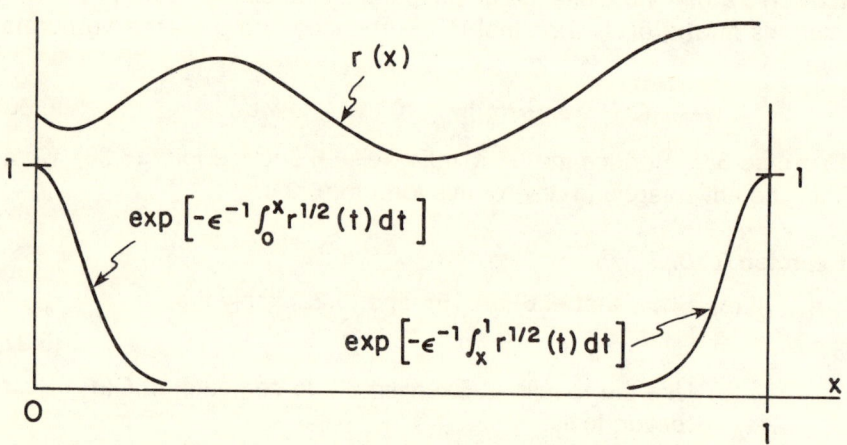

Fig. 6.3. The WKB approximation to the exponential decaying solutions of Eq. (6.33).

geneous solution is still a homogeneous solution, we multiply the WKB approximation to y_+ by $\exp(-R/\epsilon)$, where

$$R = \int_0^1 r^{1/2}(t)dt . \tag{6.35}$$

Since

$$\int_0^1 (\ \ )dt - \int_0^x (\ \ )dt = \int_x^1 (\ \ )dt , \tag{6.36}$$

it follows that in place of $(6.34)_+$, we may write

$$\tilde{y}_+ \sim r^{-1/4}(x)\exp[-\epsilon^{-1} \int_x^1 r^{1/2}(t)dt]. \tag{6.37}$$

The exponential factor now decays rapidly from 1 to 0 as x *decreases* from 1. See Fig. 6.3. The WKB approximation to the general solution of (6.33) is therefore

$$y \sim Ay_- + B\tilde{y}_+ , \tag{6.38}$$

where A and B are arbitrary constants. In y_- and $\tilde{y}_+$ we have another example of boundary layers—functions that decay rapidly to zero as we move from the boundary of a domain into its interior.

A further simplification of the WKB approximation to decaying solutions is possible because the behavior of $\int_0^x r^{1/2}(t)dt$ and $\int_x^1 r^{1/2}(t)dt$ is important only near $x=0$ and $x=1$, respectively. This is clear from Fig. 6.3: away from the end-points of the domain $(0,1)$, the exponentials are essentially zero.

To allow for the possibility that $r(x)$ is singular (but integrable) at the end-points, let us assume that

$$r(x) = \begin{cases} x^m[a_0 + O(x)], m > -2 , 0 < x \leqslant 1/2 \\ \\ (1-x)^n[b_0 + O(1-x)], n > -2, 1/2 \leqslant x < 1 \end{cases} \tag{6.39}$$

Then

$$\int_0^x r^{1/2}(t)dt = \alpha x^p[1 + O(x)] \tag{6.40}$$

$$\int_x^1 r^{1/2}(t)dt = \beta(1-x)^q[1 + O(1-x)] , \tag{6.41}$$

where

$$p = \frac{m+2}{2} > 0 , \alpha = \sqrt{a_0} /p \tag{6.42}$$

$$q = \frac{n+2}{2} > 0 , \beta = \sqrt{b_0} /q . \tag{6.43}$$

It follows that, asymptotically,

$$y \sim Ax^{-m/4}\exp[-\alpha x^p/\epsilon] + B(1-x)^{-n/4}\exp[-\beta(1-x)^q/\epsilon] , \qquad (6.44)$$

where the constants $a_0^{-1/4}$ and $b_0^{-1/4}$ coming from (6.39) have been absorbed into the arbitrary constants A and B.

The DE (6.33) can not arise in an EVP because its solutions do not oscillate. Typically, the auxiliary conditions associated with (6.33) are *non-homogeneous*, for example BC's such as

$$\lim_{x\to 0+} x^{m/4}y = a , \lim_{x\to 1-} (1-x)^{n/4}y = b . \qquad (6.45)$$

Substituting (6.44) into (6.45), we obtain

$$A \sim a , B \sim b , \qquad (6.46)$$

since $\exp(-\alpha/\epsilon)\sim 0$ and $\exp(-\beta/\epsilon)\sim 0$ as $\epsilon\to 0$. (Such exponential terms, in fact, go to zero faster than any power of ϵ, and are therefore said to be *transcendentally small*.)

Exercise 6.12. Prove this last statement. Hint: Use l'Hospital's rule to show that $\lim_{\epsilon\to 0} \epsilon^p e^{-\alpha/\epsilon} \to 0$ for every p.

Exercise 6.13. Explain why two functions $f(\epsilon)$ and $g(\epsilon)$ may have identical regular expansions but still differ by terms such as $\exp(-a/\epsilon)$. (Thus there is a limit to the amount of information that can be extracted from a regular expansion.)

Exercise 6.14. Use (6.44) to obtain an asymptotic solution to the BVP

$$\epsilon^2 y'' - x(1-x)y = 0 , \qquad 0 < x < 1$$

$$\lim_{x\to 0+} x^{1/4}y = 2 , \lim_{x\to 1-} (1-x)^{1/4}y = 1.$$

Make a careful graph of the solutions for $\epsilon = .1$ and $\epsilon = .2$. Note that the approximate solutions are singular at $x = 0$ and $x = 1$. In contrast, the exact solutions must be regular at the ends of the interval.

CHAPTER VII: TRANSITION POINT PROBLEMS AND LANGER'S METHOD OF UNIFORM APPROXIMATION

A buckled drill string[1] of length L is sketched in Fig. 7.1a.

Fig. 7.1b indicates the forces and moment on a typical cross section lying a distance s from the lower end of the string. It is assumed that the loads have rotated the section through an angle $\phi(s)$. Let $\rho(s)$ denote the mass/length of the string and V_u the force at its upper end. The vertical force at any section is then

$$V(s) = V_u - \int_s^L \rho(t)dt .\tag{7.1}$$

From force and moment equilibrium and strength of materials, we have

$$-V\sin\phi = dM/ds = EId^2\phi/ds^2,\tag{7.2}$$

where E is young's modulus (an elastic parameter) and I is the moment of inertia of the cross section about a line through the centroid, perpendicular to the plane in which bending takes place. The ends of the string are assumed to be free to rotate. Hence

$$M = EId\phi/ds = 0 \text{ at } s = 0, L.\tag{7.3}$$

Linearization and Nondimensionalization. To simplify the mathematics we assume that

$$\sin\phi \approx \phi \equiv y\tag{7.4}$$

and nondimensionalize as follows:

$$s = Lx, f = V/W, \epsilon^2 = EI/L^2W,\tag{7.5}$$

[1]"Drill string" is the name of pipe used to drill oil wells. These pipe are several inches in diameter and thousands of feet long (screwed together in 20 foot sections). As the drill string can support a transverse force as well as a tension, it is actually a *beam*.

where

$$W = \int_0^L \rho(t)dt \tag{7.6}$$

is the total weight of the string. The DE (7.2) and BC's (7.3) now take the form

$$\epsilon^2 y'' - f(x)y = 0 , \ 0<x<1 , \ y'(0) = y'(1) = 0. \tag{7.7}$$

The drill string is under tension as it is lowered. However, after its end has hit bottom, the lower part of the string will be in compression while its upper part will remain in tension so long as $V_u > 0$. The dimensionless tension $f(x)$ will have the general shape sketched in Fig. 7.2. It is convenient to shift the independent variable so that $f(0)=0$. We shall therefore study the equation

$$\epsilon^2 y'' - f(x)y = 0 , \ -a < x < b , \tag{7.8}$$

subject to certain BC's at $x = -a$ and $x = b$ which are of no concern

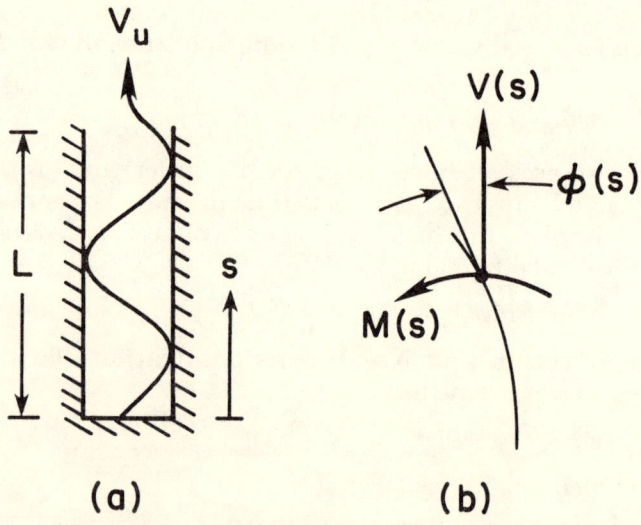

(a)   (b)

Fig. 7.1. (a) A drill string as it appears in the well casing.
(b) Loads on a section of the string.

in the analysis that follows.

If $x \leq -\delta < 0$, $f(x) < 0$, so, from (6.19), the solution of (7.8) is of the form

$$Y_{\text{osc}} \sim |f(x)|^{-1/4}\{C_1\cos[\epsilon^{-1}F(x)] + C_2\sin[\epsilon^{-1}F(x)]\}, \qquad (7.9)$$

where

$$F(x) = \int_0^x |f(t)|^{1/2}dt . \qquad (7.10)$$

If $0 < \delta \leq x$, $f(x) > 0$, so, by (6.34), the solution of (7.8) is of the form

$$Y_{\text{mon}} \sim |f(x)|^{-1/4}\{C_3\exp[-\epsilon^{-1}F(x)] + C_4\exp[\epsilon^{-1}F(x)]\}, \qquad (7.11)$$

However, in a neighborhood of $x = 0$ neither solution can be valid because as $x \to 0$, $|f(x)|^{-1/4} \to \infty$ [even though $x = 0$ is *not* a singular point of (7.8).] Nonetheless, as the solution of (7.8) must be continuous on $(-a, b)$, we can infer Fig. 7.3.

The connection problem is to express C_3 and C_4 in (7.11) as a linear combination of C_1 and C_2 in (7.9) (or vice-versa). This problem must be solved before we can solve a BVP. There are two major ways to approach the connection problem, typified by methods proposed by Jefferys and by Langer. Jefferys finds an approximate equation which he solves exactly, whereas Langer introduces a change of variable and solves

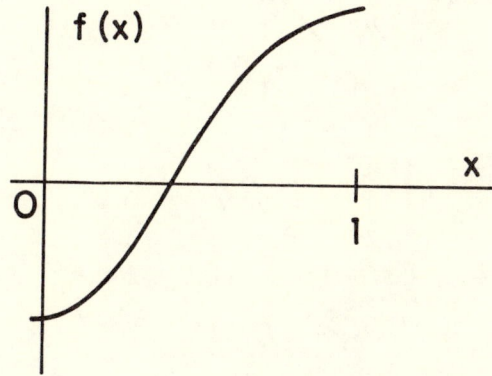

Fig. 7.2. Dimensionless tension in a drill string.

the resulting exact equation approximately.

Jefferys' Method (1923). Near $x=0$, (7.8) looks like

$$\epsilon^2 Y'' - xf'(0)Y = 0 ,\tag{7.12}$$

provided that $f'(0) \neq 0$, as we shall assume henceforth. This is a Bessel equation. We expect (hope!) that in some way, as $\epsilon \to 0$,

$$Y(x) \sim Y_{mon} , \; x>0\tag{7.13}$$

$$Y(x) \sim Y_{osc} , \; x <0.\tag{7.14}$$

The parameter ϵ as well as the constant $f'(0)$ may be scaled out of (7.12) by setting

$$x = [f'(0)]^{-1/3}\epsilon^{2/3}z.\tag{7.15}$$

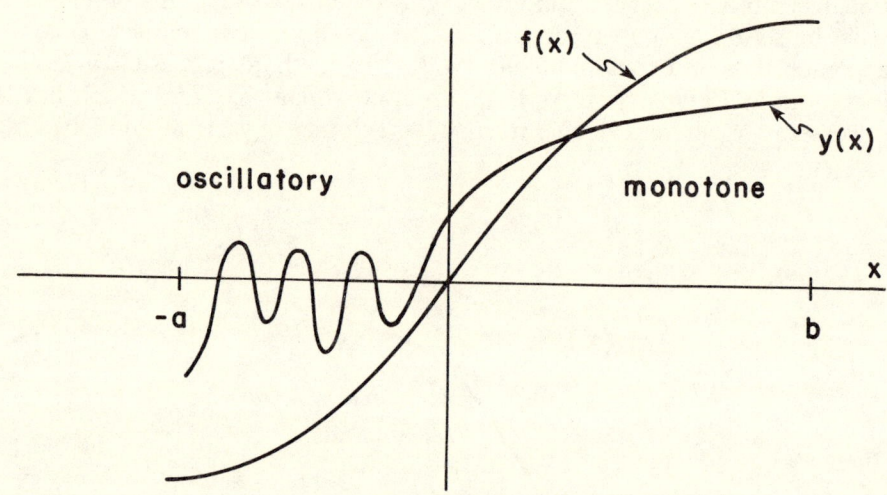

Fig. 7.3. Sketch of $f(x)$ and the solution of (7.8).

This reduces (7.12) to the *Airy equation,*

$$\frac{d^2Y}{dz^2} - zY = 0. \tag{7.16}$$

Facts About the General Solution of Airy's Equation. As $z = \infty$ is the only singular point of (7.16), this DE has two linearly independent solution analytic everywhere. These are usually denoted by $\text{Ai}(z)$ and $\text{Bi}(z)$, so that the general solution of (7.16) takes the form

$$Y(z) = C_5\text{Ai}(z) + C_6\text{Bi}(z). \tag{7.17}$$

Properties and values of $\text{Ai}(z)$ and $\text{Bi}(z)$ are given on pages 446-452 and 475-478 of the *Handbook of Mathematical Functions,* Abramowitz and Stegun, editors, U.S. Government Printing Office, 1964.

Graphs of these functions are given in Fig. 7.4.

As $z \to \infty$,

$$\text{Ai}(z) \sim \frac{\exp[-(2/3)z^{3/2}]}{2\pi^{1/2}z^{1/4}} \tag{7.18}$$

$$\text{Bi}(z) \sim \frac{\exp[(2/3)z^{3/2}]}{\pi^{1/2}z^{1/4}} \tag{7.19}$$

$$\text{Ai}(-z) \sim \frac{1}{\pi^{1/2}z^{1/4}} \sin\left\{\frac{2}{3}z^{3/2} + \frac{\pi}{4}\right\} \tag{7.20}$$

$$\text{Bi}(-z) \sim \frac{1}{\pi^{1/2}z^{1/4}} \cos\left\{\frac{2}{3}z^{3/2} + \frac{\pi}{4}\right\}. \tag{7.21}$$

Matching via an Intermediate Variable. To apply Jefferys' method, we express both z and x in terms of an *intermediate variable* η such that if we hold η fixed and let $\epsilon \to 0$, then $z \to \infty$ while $x \to 0$. To achieve this let

$$z = K\epsilon^\alpha\eta \, , \, x = \epsilon^\beta\eta \, , \, K \, , -\alpha,\beta{>}0 \, , \tag{7.22}$$

where K, α, and β are constants to be determined. Substituting (7.22) into (7.15), we get

$$\epsilon^\beta\eta = K[f'(0)]^{-1/3}\epsilon^{2/3+\alpha}\eta \, . \tag{7.23}$$

If (7.23) is to hold for η fixed and all ϵ, we must set

$$-\alpha + \beta = 2/3 \, , \, K = [f'(0)]^{1/3} \tag{7.24}$$

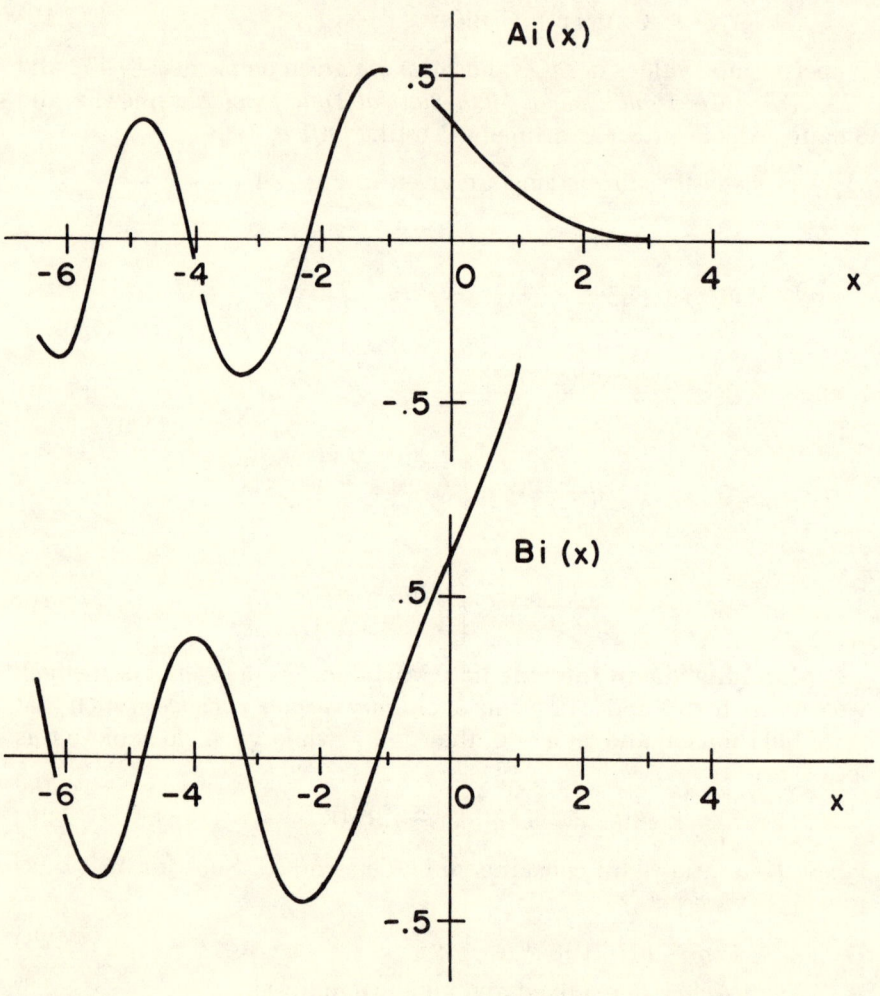

Fig. 7.4. Graphs of Ai and Bi.

Any choice of α and β satisfying (7.24) will do; for symmetry, we set $-\alpha = \beta = 1/3$. Then from (7.22),

$$z = [f'(0)]^{1/3}\epsilon^{-1/3}\eta, \ x = \epsilon^{1/3}\eta. \tag{7.25}$$

We first match (7.11) and (7.17). In this case $\eta > 0$, so that from (7.18), (7.19), and (7.25),

$$Y(z) = Y[f'(0)^{1/3}\epsilon^{-1/3}\eta]$$

$$= C_5 \text{Ai}[f'(0)^{1/3}\epsilon^{-1/3}\eta] + C_6 \text{Bi}[\cdots] \tag{7.26}$$

$$\sim C_5 \left\{ \frac{\exp[-(2/3)f'(0)^{1/2}\epsilon^{-1/2}\eta^{3/2}]}{2\pi^{1/2}f'(0)^{1/12}\epsilon^{-1/12}\eta^{1/4}} \right\} + C_6\{\cdots\},$$

while from (7.11) and (7.25),

$$Y_{mon}(x) = Y_{mon}(\epsilon^{1/3}\eta)$$

$$= C_3|f(\epsilon^{1/3}\eta)|^{-1/4}\exp[-\epsilon^{-1}F(\epsilon^{1/3}\eta)] \tag{7.27}$$

$$+ C_4|f(\epsilon^{1/3}\eta)|^{-1/4}\exp[\epsilon^{-1}F(\epsilon^{1/3}\eta)].$$

Now by Taylor's theorem,

$$f(x) = f'(0)x + O(x^2). \tag{7.28}$$

Hence

$$|f(x)|^{1/2} = f'(0)^{1/2}|x|^{1/2} + O(x^{3/2}), \tag{7.29}$$

so that from (7.10),

$$F(x) = \frac{2}{3}f'(0)^{1/2}x^{3/2} + O(x^{5/2}), \ x > 0$$

$$\sim \frac{2}{3}f'(0)^{1/2}\epsilon^{1/2}\eta^{3/2}. \tag{7.30}$$

Thus

$$Y_{mon}(\epsilon^{1/3}\eta) \sim C_3 \left\{ \frac{\exp[-(2/3)f(0)^{1/2}\epsilon^{-1/2}\eta^{3/2} + \cdots]}{f'(0)^{1/4}\epsilon^{1/12}\eta^{1/4}} \right\} + C_4\{\cdots\}. \tag{7.31}$$

If (7.26) and (7.31) are to agree, then

$$\frac{C_3}{f'(0)^{1/4}\epsilon^{1/12}} = \frac{C_5}{2\pi^{1/2}f'(0)^{1/12}\epsilon^{-1/12}} \tag{7.32}$$

or

$$C_3 = \frac{[\epsilon f'(0)]^{1/6}}{2\pi^{1/2}} C_5.$$
(7.33)

Exercise 7.1. Find the relation between C_4 and C_6.

Exercise 7.2. Show that (7.14) implies

$$\begin{Bmatrix} C_1 \\ C_2 \end{Bmatrix} = \frac{[\epsilon f'(0)]^{1/6}}{(2\pi)^{1/2}} \begin{Bmatrix} C_5 + C_6 \\ C_5 - C_6 \end{Bmatrix}.$$
(7.34)

Exercise 7.3. Obtain asymptotic solutions to the EVP's

  (a) $\epsilon^2 y'' - [\sin(\pi x/2)]y = 0$, $y(-1) = y(1) = 0$.

  (b) $\epsilon^2 y'' - (\tanh x)y = 0$, $y'(-1) = y'(1) = 0$.

Langer's Method. Our treatment of (7.8) for $\epsilon^2 < 1$ required us to construct approximate solutions on the three domains $-a < x < -\delta$, $-\delta < x < \delta$, and $\delta < x < b$ and then to relate the various constants of integration by a matching procedure. The approximate solution on $(-\delta, \delta)$, (7.17), was a linear combination of the well-studied Airy functions, $\text{Ai}(z)$ and $\text{Bi}(z)$.

It was Langer's idea to attempt to describe the behavior of the solution of (7.8) over the *entire* domain $-a < x < b$ in terms of Ai and Bi. His method consists of introducing new dependent and independent variables such that, except for small terms, (7.8) takes the form of (7.16). Langer's work was initiated in the early 1930's and completed in the late 1940's. The two-scale method, which first appeared about 10 years later, turns out to be an ideal way of arriving at Langer's results.

Recall that to study the behavior of (7.8) near the transition point $x = 0$, we approximated $f(x)$ by $xf'(0)$ and then introduced the new independent variable

$$z = \epsilon^{-2/3}[f'(0)]^{1/3}x .$$
(7.35)

Compared to x, z is a "fast variable" (in Cole and Kevorkian's terminology), since a change in x of $O(\epsilon^{2/3})$ produces a change in z of $O(1)$. The results of the preceding section and the form of (7.35) suggests that we attempt to express the exact solution of (7.8) as a function of

$$\xi = \epsilon^{-2/3}g(x),$$
(7.36)

and x. To fix $g(x)$, we shall require that, to a first approximation, we obtain a DE in terms of the fast variable ξ alone, identical to the Airy equation. Finally, we shall make $g(0) = 0$ so that the transition point occurs at $\xi = 0$.

Thus with $\beta = \epsilon^{2/3}$ we assume that

$$y = y(x,\xi,\beta). \tag{7.37}$$

Then

$$y' = \frac{dy}{dx} = \frac{\partial y}{\partial x} + \frac{\partial y}{\partial \xi}\frac{d\xi}{dx}$$

$$= y_x + \beta^{-1}g'y_\xi \tag{7.38}$$

$$y'' = y_{xx} + 2\beta^{-1}g'y_{x\xi} + \beta^{-1}g''y_\xi + \beta^{-2}g'^2y_{\xi\xi}. \tag{7.39}$$

Inserting (7.39) into (7.8), and multiplying through by $\xi = \beta^{-1}g$, we obtain

$$gg'^2y_{\xi\xi} + \beta g(2g'y_{x\xi} + g''y_\xi) + \beta^2 gy_{xx} - \xi f(x)y = 0. \tag{7.40}$$

To make (7.40) take the form of (7.16) as $\beta \to 0$, we set

$$gg'^2 = f. \tag{7.41}$$

Upon division by (7.41), (7.40) reduces to

$$y_{\xi\xi} + \beta[(2/g')y_{x\xi} + (g''/g'^2)y_\xi] + \beta^2(1/g'^2)y_{xx} - \xi y = 0. \tag{7.42}$$

Since g has the same sign as f and $g(0)=0$, the solution of (7.41) is

$$g(x) = \text{sgn}(x)[(3/2)\int_0^x |f(t)|^{1/2}dt]^{2/3}. \tag{7.43}$$

Now assume an expansion for y of the form

$$y(\xi,x,\beta) = \overset{0}{y}(\xi,x) + \beta\overset{1}{y}(\xi,x) + \cdots. \tag{7.44}$$

Substituting (7.44) into (7.42) and equating to zero the coefficients of successive powers of β, we obtain the sequence of DE's:

$$\overset{0}{y}_{\xi\xi} - \xi\overset{0}{y} = 0 \tag{7.45}_0$$

$$\overset{1}{y}_{\xi\xi} - \xi\overset{1}{y} = -(2/g')\overset{0}{y}_{\xi x} - (g''/g'^2)\overset{0}{y}_\xi, \tag{7.45}_1$$

etc., the first of which has the general solution

$$\overset{0}{y}(\xi,x) = C_0(x)\text{Ai}(\xi) + D_0(x)\text{Bi}(\xi). \tag{7.46}$$

To determine the functions $C_0(x)$ and $D_0(x)$ we must consider (7.45)$_1$.

For simplicity we determine $C_0(x)$ only; $D_0(x)$ may be found in a strictly analogous fashion. Inserting (7.46) with $D_0=0$ into the right side of $(7.45)_1$, we have

$$\tfrac{1}{y}_{\xi\xi} - \xi\tfrac{1}{y} = -[(2/g')C_0' + (g''/g'^2)C_0]\text{Ai}'(\xi). \tag{7.47}$$

Observe that if $A(\xi)$ is a solution of Airy's DE, then

$$(\xi A)'' - \xi(\xi A) = 2A'. \tag{7.48}$$

Thus the right side of (7.47) will produce a particular solution of the form $F(x)\xi\text{Ai}(\xi)$ unless the coefficient of Ai' is zero. To avoid such "secular" terms in $\tfrac{1}{y}(\xi,x)$ we choose $C_0(x)$ so that

$$(2/g')C_0' + (g''/g'^2)C_0 = 0. \tag{7.49}$$

This equation has the solution

$$C_0(x) = \frac{\gamma_0}{[g'(x)]^{1/2}}$$

$$= \gamma_0(g/f)^{1/4}, \tag{7.50}$$

where γ_0 is an arbitrary constant. Hence, from (7.36), (7.43), and (7.50),

$$y = \gamma_0[g(x)/f(x)]^{1/4}\text{Ai}[\epsilon^{-2/3}g(x)] + O(\epsilon^{2/3}), \tag{7.51}$$

where $g(x)$ is given by (7.43). It may be shown that the error estimate is rigorous. (See Section 12.8 of "Asymptotic Methods" by F. Olver in *Handbook of Applied Mathematics*, Ed. C. Pearson, Van Nostrand, 1974).

As $x \to +\infty$, we have, using (7.18) and (7.43),

$$y \sim \frac{\gamma_0 \, \epsilon^{1/6}}{2\pi^{1/2}[f(x)]^{1/4}}\exp\{-\epsilon^{-1}\int_0^x[f(t)]^{1/2}dt\}, \tag{7.52}$$

which agrees with (7.17), while as $x \to -\infty$, we have, using (7.20),

$$y \sim \frac{\gamma_0 \, \epsilon^{1/6}}{\pi^{1/2}|f(x)|^{1/4}}\sin[\epsilon^{-1}\int_0^x|f(t)|^{1/2}dt + \pi/4], \tag{7.53}$$

which agrees with (7.9).

Exercise 7.4. Compute $D_0(x)$ in (7.46).

Exercise 7.5. Carry the computations far enough to determine $\tfrac{1}{y}(x,\xi)$. Is there any evidence that a more general fast variable of the form $\xi = \beta^{-1}g(x,\beta)$ should have been introduced?

Exercise 7.6. Consider the EVP

$$y'' + \omega^2(\sin x)y = 0 \ , \ y(0) = y(1) = 0.$$

The WKB method yields (6.22) as a first approximation to the nth natural frequency ω_n, but fails, as you saw in Exercise 6.10(c), to give a correction. Use Langer's method to determine this improved approximation.

CHAPTER VIII:
INTRODUCTION TO BOUNDARY LAYER THEORY

In Chapter III we considered a simple though typical boundary value problem, namely,

$$\epsilon y'' + y' + y = 0 \, , \, 0 < x < 1 \, , \, y(0) = 0 \, , \, y(1) = 1. \quad (8.1)$$

The approximate solution given there was

$$\bar{y} = e(e^{-x} - e^{-x/\epsilon}), \qquad (8.2)$$

a sum of two functions whose graphs are indicated by the dashed lines in Fig. 8.1. The solution of (8.1) exhibits a boundary layer of width $O(\epsilon)$ at the left end of the interval $[0,1]$.

The aim of this chapter is to develop ways to obtain, systematically, approximate solutions to BVP's that exhibit boundary layers. To lay the footings for later perturbation expansions we now introduce, via a few simple examples, a method of obtaining *first-approximation* solutions.

The Boundary Layer Method. The first step is to see what happens if we set the small parameter in our problem to zero. For the problem at hand, (8.1), we obtain, upon setting $\epsilon = 0$, the reduced DE

$$Y' + Y = 0 \, , \qquad (8.3)$$

whose general solution is

$$Y = Ce^{-x}. \qquad (8.4)$$

As (8.4) contains only one arbitrary constant, C, we cannot satisfy both boundary conditions in (8.1). Looking at the graph of $\bar{y}$ in Fig. 8.1, which we expect to be a reasonable picture of the solution of (8.1) if ϵ is small, we see that the solution changes very rapidly near the left boundary. To exploit this observation mathematically, we introduce a new, *boundary layer variable*,

$$\xi = x/\epsilon, \tag{8.5}$$

whereupon the DE in (8.1), when multiplied by ϵ, becomes

$$\frac{d^2y}{d\xi^2} + \frac{dy}{d\xi} + \epsilon y = 0. \tag{8.6}$$

Setting $\epsilon = 0$, we obtain

$$\frac{d^2y_{BL}}{d\xi^2} + \frac{dy_{BL}}{d\xi} = 0, \tag{8.7}$$

whose solution is

$$y_{BL} = A + Be^{-\xi}. \tag{8.8}$$

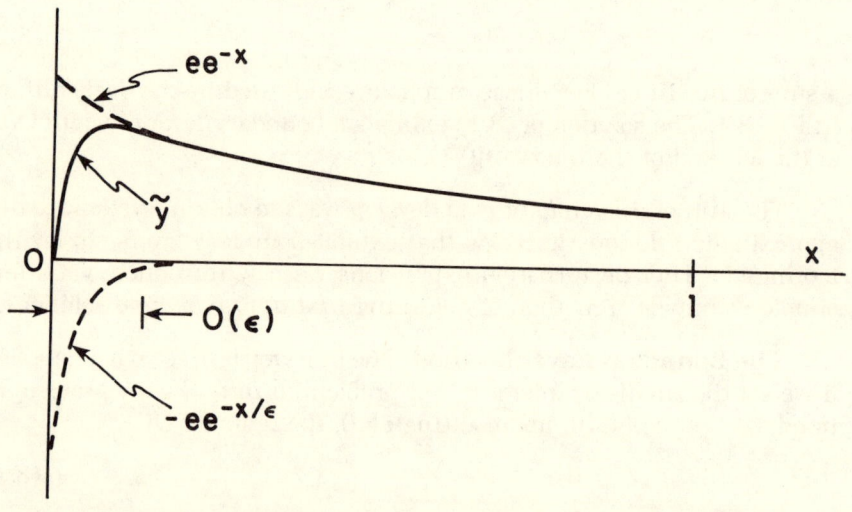

Fig. 8.1. The functions ee^{-x} and $-ee^{-x/\epsilon}$ and their sum $\tilde{y}$.

From (8.8), we see that the boundary layer will be on the left because the term $e^{-\xi}$ goes to zero as we move away from the left boundary. [If there had been a negative sign between the first two terms in (8.6), the boundary layer would have been on the right.]

We must now try to combine (8.4) and (8.8) to make an approximate solution of (8.1). As the boundary layer is on the left, we may find C by forcing $Y(1) = 1$. Thus

$$Y = ee^{-x}. \tag{8.9}$$

This is called an *outer* or *interior solution* (meaning that it is valid in the interior of the region, away from the boundary layer). The solution (8.8) must therefore meet the boundary condition on the left, that is,

$$y_{BL} = A(1 - e^{-\xi}). \tag{8.10}$$

The constant A is determined by making (8.9) and (8.10) match at the edge of the boundary layer. Heuristically, we want the limit of the outer solution as x approaches the boundary layer to equal the limit of the boundary layer solution as ξ approaches the outer region. In symbols, we want

$$\lim_{x \to 0} Y = \lim_{\xi \to \infty} y_{BL}, \tag{8.11}$$

which, by (8.9) and (8.10) yields

$$A = e. \tag{8.12}$$

In certain problems, matching is more complicated and must be carried out in a more formal way by introducing an intermediate variable—call it z—similar to that used in Jefferys' method for the turning point problem. To demonstrate the method with (8.9) and (8.10) we set

$$x = \epsilon^{\alpha} z, \; \xi = \epsilon^{\beta} z. \tag{8.13}$$

We want α to be positive and β to be negative so that, for fixed z, x will go to zero and ξ will go to infinity as ϵ goes to zero. Because $\xi = x/\epsilon$, we have

$$\beta = \alpha - 1. \tag{8.14}$$

We take $\alpha = \frac{1}{2}$, $\beta = -\frac{1}{2}$, substitute (8.9) and (8.10) into (8.11) and get

$$\lim ee^{-\sqrt{\epsilon} z} = \lim A(1 - e^{-z/\sqrt{\epsilon}}), \tag{8.15}[1]$$

which implies that

$$A = e. \tag{8.16}$$

---

[1]Recall that lim is short for "the limit as $\epsilon \to 0$."

Our approximate solution is now given by two pieces:

$$y \approx \begin{cases} e(1 - e^{-x/\epsilon}) \,, & x < O(\epsilon) \\ ee^{-x} \,, & x > O(\epsilon) \end{cases} \,, \tag{8.17}$$

Equation (8.17) is awkward because we must decide when to stop using one expression and to start using the other. A uniformly valid solution may be found by adding the two solutions and subtracting the part which is common to both. This common part is simply the common limit given by (8.15). Thus

$$\begin{aligned} y_{\text{uniform}} &= y_{BL} + Y - \text{common part} \\ &= e(1 - e^{-x/\epsilon}) + ee^{-x} - e \\ &= e(e^{-x} - e^{-x/\epsilon}). \end{aligned} \tag{8.19}$$

As a slight modification of our original problem, (8.1), let us consider

$$\epsilon y'' - y' + y = 0 \,, \, 0 < x < 1 \,, \, y(0) = 0 \,, \, y(1) = 1. \tag{8.20}$$

The first thing we notice is that a boundary layer will form on the right because the solution of $\epsilon y'' - y' \approx 0$ decreases as x decreases. The reduced equation,

$$Y' - Y = 0 \,, \, Y(0) = 0, \tag{8.21}$$

has the solution

$$Y = 0. \tag{8.22}$$

Introducing the boundary layer coordinate

$$\xi = (x - 1)/\epsilon, \tag{8.23}$$

which is chosen to vanish at the right boundary ($x = 1$), we obtain the boundary layer equation

$$\frac{d^2 y_{BL}}{d\xi^2} - \frac{dy_{BL}}{d\xi} = 0 \,, \, y_{BL}(0) = 1 \tag{8.24}$$

with solution

$$y_{BL} = 1 - B(1 - e^{\xi}). \tag{8.25}$$

If we match (8.25), as we move away from the boundary layer, to (8.22) as we move toward the boundary layer, we find that $B = 1$. Therefore,

$$y_{\text{uniform}} = e^{(1-x)/\epsilon}. \tag{8.26}$$

Here, the common part, as well as the outer solution, is zero.

An interesting variation on this type of problem is one which does not have a boundary layer at either end. Consider

$$\epsilon y'' + xy' + x^2 y = 0 , \quad -1 < x < 1 \tag{8.27}$$

$$y(-1) = \alpha , \; y(1) = \beta . \tag{8.28}$$

There is no boundary layer on the left $(x = -1)$ because the sign between the first and second term of the DE is negative, but neither is there a boundary layer on the right $(x = 1)$ because the sign there is positive. The reduced equation

$$xY' + x^2 Y = 0 \tag{8.29}$$

has the solution

$$Y = Ce^{-\frac{1}{2}x^2}. \tag{8.30}$$

We can meet both boundary conditions in (8.28) by choosing C differently for $x < 0$ and $x > 0$. Thus

$$Y = \begin{cases} \alpha e^{\frac{1}{2}} e^{-\frac{1}{2}x^2} , & x < 0 \\ \beta e^{\frac{1}{2}} e^{-\frac{1}{2}x^2} , & x > 0 \end{cases} . \tag{8.31}$$

A sketch of (8.31) is shown in Fig. 8.2 for α and β positive.

It is clear from Fig. 8.2 that the first and second derivatives of the exact solution of (8.27) and (8.28) must be large near $x = 0$. To study this behavior we introduce the new variable

$$\xi = x/\sqrt{\epsilon}, \tag{8.32}$$

in terms of which (8.27) reads

$$\frac{d^2 y}{d\xi^2} + \xi \frac{dy}{d\xi} + \epsilon \xi^2 y = 0. \tag{8.33}$$

Setting $\epsilon = 0$ produces a boundary layer equation with solution

$$y_{BL} = A + B \int_{\xi}^{\infty} e^{-1/2t^2} dt . \tag{8.34}$$

[For simplicity let $J(\xi) = \int_{\xi}^{\infty} e^{-1/2t^2} dt$. Then $2J(0) = J(-\infty) = \sqrt{2\pi}$]. Using the quick matching procedure

$$\lim_{\xi \to \infty} y_{BL} = \lim_{x \to 0+} Y \tag{8.35}$$

$$\lim_{\xi \to -\infty} y_{BL} = \lim_{x \to 0-} Y ,$$

we obtain equations for A and B:

$$A = \beta e^{1/2}$$

$$A + \sqrt{2\pi}\, B = \alpha e^{1/2}.$$

(8.36)

Solving (8.36) for A and B, we produce the following uniform solution:

$$y_{\text{uniform}} = \begin{cases} \beta e^{1/2(1-x^2)} + e^{1/2}(\alpha - \beta)J(x/\sqrt{\epsilon})/\sqrt{2\pi}, & x > 0 \\ \alpha e^{1/2(1-x^2)} + e^{1/2}(\alpha - \beta)[J(x/\sqrt{\epsilon})/\sqrt{2\pi} - 1], & x < 0 \end{cases}$$

(8.37)

At $x = 0$ each expression on the right of (8.37) has the value $(1/2)(\alpha + \beta)e^{1/2}$.

The reader should notice that the first two examples in this chapter were ones that we could have solved exactly whereas the last example would have been quite difficult to do exactly. A more complicated example, illustrating matching in a problem of physical origin, is given in Chapter IX.

The Two-Scale Method. If we examine (8.2), we see two scales; a slow one x and a fast one $\xi = x/\epsilon$. If these are introduced formally into

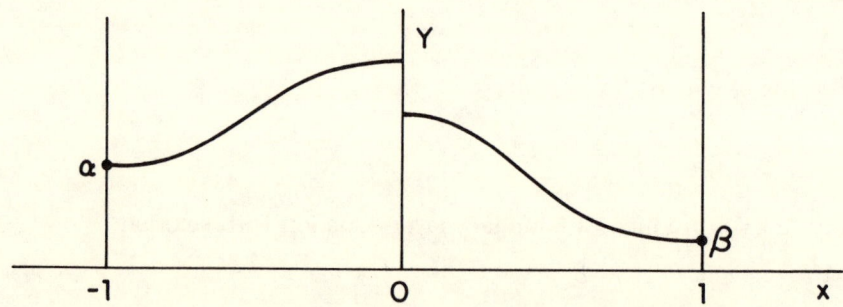

Fig. 8.2. Solution of the reduced DE (8.29) meeting the BC's (8.28).

(8.1), as was done in Chapter V, we obtain

$$\epsilon\left\{\frac{\partial^2 y}{\partial x^2} + \frac{2}{\epsilon}\frac{\partial^2 y}{\partial x \partial \xi} + \frac{1}{\epsilon^2}\frac{\partial^2 y}{\partial \xi^2}\right\} + \frac{\partial y}{\partial x} + \frac{1}{\epsilon}\frac{\partial y}{\partial \xi} + y = 0 \tag{8.38}$$

$$y(0,0,\epsilon) = 0 \ , \ y(1,\epsilon^{-1},\epsilon) = 1. \tag{8.39}$$

Clearing the ϵ and setting

$$y(x,\xi,t) = \overset{0}{y}(x,\xi) + \epsilon\overset{1}{y}(x,\xi) + \cdots , \tag{8.40}$$

we have for the coefficients of ϵ^0 and ϵ^1,

$$\overset{0}{y}_{\xi\xi} + \overset{0}{y}_{\xi} = 0 \tag{8.41$_0$}$$

$$\overset{1}{y}_{\xi\xi} + \overset{1}{y}_{\xi} = -2\overset{0}{y}_{x\xi} - \overset{0}{y}_x - \overset{0}{y}. \tag{8.41$_1$}$$

The solution of $(8.41)_0$ is

$$\overset{0}{y} = A(x) + B(x)e^{-\xi}, \tag{8.42}$$

which reduces $(8.41)_1$ to

$$\overset{1}{y}_{\xi\xi} + \overset{1}{y}_{\xi} = (B' - B)e^{-\xi} - (A' + A), \tag{8.43}$$

To suppress terms like $\xi e^{-\xi}$ and ξ in the particular solution of (8.43), we set the coefficient of $e^{-\xi}$ and ξ^0 to zero. This yields

$$A' + A = 0 \ , \ B' - B = 0. \tag{8.44}$$

The boundary conditions (8.39) imply that

$$A(0) + B(0) = 0 \ , \ A(1) = 1. \tag{8.45}$$

Solving (8.44) and (8.45), we obtain from (8.42)

$$\overset{0}{y} = e(e^{-x} - e^x e^{-x/\epsilon}). \tag{8.46}$$

Note that we assumed the boundary layer was on the left when we set $\xi = x/\epsilon$ because $\xi = O(1)$ in a neighborhood of $x = 0$. If the boundary layer had been on the right, we would have failed.

The Two-Scale Method for Variable Coefficient DE's. Consider the variable coefficient DE and BC's

$$\epsilon y'' + a(x)y' + b(x)y = 0 \ , \ 0 < x < 1 \ , \ y(0) = 0 \ , \ y(1) = 1, \tag{8.47}$$

where $a(x)$ is continuously differentiable and *positive* on $(0,1)$. If ϵ is small, $a(x)$ changes little over an interval of length $O(\epsilon)$. Applying the same heuristic reasoning as used in Chapter III, we might expect a uniformly valid, first approximation solution of the form

$$\bar{y} = -Y(0)e^{-(\bar{a}/\epsilon)x} + Y(x), \tag{8.48}$$

where $\bar{a}$ is some average value of a taken over a small interval near $x=0$.

To obtain more quantitative information, we assume that the solution of (8.47) depends on a slow variable x and a fast variable

$$\xi = \epsilon^{-1}g(x) , \ g(0)=0 \tag{8.49}$$

and proceed formally:

$$y = y(x,\xi,\epsilon)$$

$$y' = y_x + y_\xi \epsilon^{-1}g' \tag{8.50}$$

$$y'' = y_{xx} + 2y_{x\xi}\epsilon^{-1}g' + y_\xi \epsilon^{-1}g'' + y_{\xi\xi}\epsilon^{-2}g'^2.$$

Substituting (8.49) and (8.50) into the BVP (8.47), we get, after multiplying by ϵ all equations[2]

$$g'^2 y_{\xi\xi} + ag'y_\xi + \epsilon(2y_{x\xi}g' + y_\xi g'' + ay_x + by) + \epsilon^2 y_{xx} = 0 \ (8.51)$$

$$y(0,0,\epsilon)=0 , \ y[1,\epsilon^{-1}g(1),\epsilon] = 1.$$

We now assume an expansion for y of the form

$$y(x,\xi,\epsilon) = \overset{0}{y}(x,\xi) + \epsilon\overset{1}{y}(x,\xi) + \cdots, \tag{8.52}$$

substitute this into (8.51), and equate to zero coefficients of successive powers of ϵ. This yields the sequence of BVP's

$$g'^2 \overset{0}{y}_{\xi\xi} + ag'\overset{0}{y}_\xi = 0 , \ \overset{0}{y}(0,0) = 0 , \ \overset{0}{y}[1,\epsilon^{-1}g(1)]=1 \tag{8.53$_0$}$$

$$g'^2 \overset{1}{y}_{\xi\xi} + ag'\overset{1}{y}_\xi + 2\overset{0}{y}_{x\xi}g' + \overset{0}{y}_\xi g'' + a\overset{0}{y}_x + b\overset{0}{y} = 0$$

$$\overset{1}{y}(0,0) = 0 , \ \overset{1}{y}[1,\epsilon^{-1}g(1)] = 0, \tag{8.53$_1$}$$

etc.

To make (8.53)$_0$ as simple as possible and to meet the initial condition in (8.49), we set

$$g(x) = \int_0^x a(t)dt , \ i.e., \ g' = a. \tag{8.54}$$

---

[2]We could have, by a change of the dependent and/or independent variable, reduced the DE in (8.47) to one of the normal forms discussed in Chapter VI. We do not because the success of boundary layer methods does not depend on this reduction which may not exist for nonlinear or higher order DE's.

The PDE in $(8.53)_0$ then reduces to

$$\overset{0}{y}_{\xi\xi} + \overset{0}{y}_\xi = 0, \tag{8.55}$$

which has the solution

$$\overset{0}{y}(x,\xi) = A_0(x) + B_0(x)e^{-\xi}. \tag{8.56}$$

The first BC in $(8.53)_0$ yields

$$A_0(0) + B_0(0) = 0. \tag{8.57}$$

At $x = 1$, $e^{-\xi} = \exp[-\epsilon^{-1}\int_0^1 a(t)dt] \leq Ce^{-K/\epsilon}$, C,K constant. Thus $e^{-\xi}$ is a *transcendentally small term* (TST) and the second BC in $(8.53)_0$ reduces to

$$A_0(1) = 1. \tag{8.58}$$

To determine $A_0(x)$ and $B_0(x)$, we must consider the form of the solution of $(8.53)_1$. With $g(x)$ given by (8.54) and $\overset{0}{y}$ by (8.56), there follows, after division by $g'^2 = a^2$,

$$\overset{1}{y}_{\xi\xi} + \overset{1}{y}_\xi = e^{-\xi}\left[\frac{B_0'}{a} + \frac{(a'-b)B_0}{a^2}\right] - \left(\frac{A_0'}{a} + \frac{bA_0}{a^2}\right). \tag{8.59}$$

To avoid "resonant" terms of the form $F(x)\xi e^{-\xi} + G(x)\xi$ in the solution for $\overset{1}{y}$, we must set to zero the coefficients of $e^{-\xi}$ and ξ^0 on the right of (8.59). In view of (8.58) and the expression in parentheses,

$$aA_0' + bA_0 = 0, \tag{8.60}$$

which has the solution

$$A_0(x) = \exp\{\int_x^1 [b(t)/a(t)]dt\}, \tag{8.61}$$

while

$$aB_0' + (a' - b)B_0 = 0 \tag{8.62}$$

has the solution

$$B_0(x) = [C_0/a(x)]\exp\{\int_0^x [b(t)/a(t)]dt\}. \tag{8.63}$$

To determine the constant C_0 in (8.63), we apply the BC (8.57). With $A_0(x)$ given by (8.61) and

$$f(x) \equiv \int_0^x [b(t)/a(t)]dt, \tag{8.64}$$

we find that

$$C_0 = -a(0)\exp[f(1)]. \tag{8.65}$$

So far, then,

$$y = \overset{0}{y}(\xi,x) + O(\epsilon)$$
$$= \exp[f(1) - f(x)] \tag{8.66}$$
$$- [a(0)/a(x)]\exp[-\epsilon^{-1}\int_0^x a(t)dt + f(x) + f(1) + O(\epsilon)].$$

Exercise 8.1. Determine $\overset{0}{y}(\xi,x)$ for the BVP

$$\epsilon y'' + (1 + x)y' + x^2 y = 0 \,, \, y(0) = 0 \,, \, y(1) = 1.$$

Exercise 8.2. Determine $\overset{0}{y}(\xi,x)$ for the BVP

$$\epsilon y'' + xy' + x^2 y = 0 \,, \, y(0) = 0, \, y(1) = 1.$$

Try the problem using the two-scale method and the boundary layer method. Which method do you prefer?

Exercise 8.3. The BVP

$$\epsilon y'' + a(x)y' = 0 \,, \, y(0) = 0 \,, \, y(1) = 1$$

can be solved exactly, since the DE is first order in y'. Compare the solution with (8.66).

Exercise 8.4. Find the expression analogous to (8.66) for the BVP

$$\epsilon y'' + a(x)y' + b(x)y = 0 \,, \, y'(0) = 1, \, y(1) = 0.$$

where $a(x)$ is differentiable and positive on $(0,1)$.

Exercise 8.5. Consider the BVP

$$L_\epsilon y \equiv \epsilon^2 y'''' - a(x)y'' + y = 0 \,, \, 0 < x < 1,$$
$$y(0) = 1 \,, \, y'(0) = 0 \,, \, y(1) = y'(1) = 0.$$

Suppose that $a(x) > 0$ on $(0,1)$ and that two independent solutions of $L_0 y = 0$ are known, call them $U(x)$ and $V(x)$. Find a uniformly valid approximation to the solution using the boundary layer method.

Exercise 8.6. Suppose that a beam under tension rests on a foundation with a modulus that varies linearly along the length of the beam. Nondimensionalizing as in Exercise 1.14, assuming no external distributed load, and considering a semi-infinite beam under the same BC's as in Exercise 5.14, we arrive at the BVP

$$[\epsilon^2 y'''' + (1 + mx)y] - y'' = 0 \,, \, 0 < x \,, \, 0 < m$$
$$y(0) = 1 \,, \, y'(0) = 0 \,, \, y(x) \,, \, y'(x) \to 0 \text{ as } x \to \infty.$$

If $0 < \epsilon << 1$, there are, as in Exercise 5.14, two disparate scales. Introduce appropriate short and long lengths, X and ξ, and determine a uniformly valid first-approximation $\overset{0}{y}(X,\xi)$ to $y(x,\epsilon)$.

Matched Asymptotic Expansions. The two-scale method can fail if closed form solutions cannot be obtained for the associated PDE's. In this case, a solution by a so-called matched asymptotic expansion procedure may sometimes work. As applied to (8.47), it rests upon the recognition that the solution has distinctly different behavior in different subintervals of $[0,1]$. In the "inner" or boundary layer region, $[0,O(\epsilon)]$, the solution is such that

$$\epsilon^2 y'' + \epsilon a(0)y' + O(\epsilon) = 0, \tag{8.67}$$

whereas in the "outer" region, $[O(\epsilon),1]$, the solution of (8.47) is such that

$$a(x)y' + b(x)y + O(\epsilon) = 0. \tag{8.68}$$

The form of (8.68) suggests that in the outer region, we assume an expansion of the form

$$y = Y(x,\epsilon) = Y_0(x) + \epsilon Y_1(x) + \cdots \tag{8.69}$$

Substituting (8.69) into (8.47) and equating to zero the coefficients of like powers of ϵ, we obtain the sequence of DE's,

$$a(x)Y_0' + b(x)Y_0 = 0 \tag{8.70}_0$$

$$a(x)Y_k' + b(x)Y_k = -Y_{k-1}'', \quad k = 1,2, \cdots, \tag{8.70}_k$$

which has the sequence of general solutions

$$Y_0(x) = C_0 \exp[-f(x)], \quad f(x) = \int_0^x (b/a)dt. \tag{8.71}_0$$

$$Y_1(x) = \left\{ C_1 + C_0 \int_0^x \frac{1}{a}\left[\left(\frac{b}{a}\right)' - \frac{b^2}{a^2}\right]dt \right\} \exp[-f(x)], \tag{8.71}_1$$

etc.,

Application of the BC $y(1)=1$ yields

$$C_0 = \exp[f(1)] \tag{8.72}_0$$

$$C_1 = \int_0^1 \frac{1}{a}\left[\frac{b^2}{a^2} - \left(\frac{b}{a}\right)'\right]dt, \tag{8.72}_1$$

etc.

In the inner region, we set

$$x = \epsilon\xi, \tag{8.73}$$

so that $\xi = O(1)$, and replace the variable coefficients in (8.47) by their

Taylor series expansions. This puts the DE into the form

$$\frac{d^2y}{d\xi^2} + (a_0 + a_1\epsilon\xi + a_2\epsilon^2\xi^2 + \cdots) \frac{dy}{d\xi}$$

$$+ \epsilon(b_0 + b_1\epsilon\xi + b_2\epsilon^2\xi^2 + \cdots)y = 0. \qquad (8.74)$$

Assuming an expansion of the form

$$y = \tilde{Y}(\xi,\epsilon) = \tilde{Y}_0(\xi) + \epsilon\tilde{Y}_1(\xi) + \cdots \qquad (8.75)$$

in the inner region, substituting this into (8.74) and equating to zero coefficients of like powers of ϵ, we obtain the sequence of DE's

$$\tilde{Y}_0'' + a_0\tilde{Y}_0' = 0 \qquad (8.76)_0$$

$$\tilde{Y}_1'' + a_0\tilde{Y}_1' = -a_1\xi\tilde{Y}_0' - b_0\tilde{Y}_0, \qquad (8.76)_1$$

etc., which has the sequence of solutions

$$\tilde{Y}_0(\xi) = B_0 + C_0e^{a_0\xi} \qquad (8.77)_0$$

$$\tilde{Y}_1(\xi) = B_1 + C_1e^{-a_0\xi} - (B_0b_0/a_0)\xi$$

$$+ C_0(b_0/a_0 - a_1/a_0 - \tfrac{1}{2}a_1\xi)\xi e^{-a_0\xi}, \qquad (8.77)_1$$

etc. Applying the BC $y(0) = 0$, we obtain

$$B_0 + C_0 = 0 \qquad (8.78)_0$$

$$B_1 + C_1 = 0, \qquad (8.78)_1$$

etc.

The Matching Procedure. To obtain additional conditions for the three sets of constants A_k, B_k and C_k, we must match the inner and outer expansions. To this end, we choose an intermediate variable η such that if η is fixed and $\epsilon \to 0$, then $\xi \to \infty$ and $x \to 0$. A convenient choice, consistent with (8.73), is

$$\xi = \beta^{-1}\eta \,, \; x = \beta\eta, \text{ where } \beta = \epsilon^{1/2}. \qquad (8.79)$$

We now express both the outer and inner solutions in terms of η:

$$Y = Y_0(\beta\eta) + \beta^2 Y_1(\beta\eta) + \cdots$$

$$= e^{f(1)} - (b_0/a_0)\beta\eta e^{f(1)} + O(\beta^2).$$

(8.80)

$$\tilde{Y} = \tilde{Y}_0(\beta^{-1}\eta) + \beta^2 \tilde{Y}_1(\beta^{-1}\eta) + \cdots$$

$$= B_0 - B_0(b_0/a_0)\beta\eta + O(\beta^2).$$

(8.81)

Comparing (8.80) with (8.81), we see that $B_0 = e^{f(1)}$. To determine B_1, it is necessary to consider the $O(\beta^2)$ – terms.

The composite expansion is defined to be

$$Y_c \equiv Y + \tilde{Y} - \text{common part.}$$

(8.82)

In the present case, the common part, to lowest order, is $e^{f(1)}$. Hence

$$Y_c = e^{f(1)-f(x)} - e^{-a_0 x/\epsilon + f(1)} + O(\epsilon).$$

(8.83)

This is essentially the same result we obtained by the two-scale method. See (8.66).

Corner or Interior Boundary Layers. In the preceding analysis, we assumed that $a(x) > 0$ on $[0,1]$. If this is not the case, *e.g.,* if $a(x)$ has a zero on $(0,1)$, *interior boundary layers* may exist. The following simple problem illustrates such a phenomenon.

$$\epsilon^2 y'' + 2xy' - 2y = 0 , \ -1 < x < 1 , \ y(-1) = A , \ y(1) = B.$$

(8.84)

[The interval $(-1,1)$ has been chosen rather than $(0,1)$ to simplify the subsequent algebra.] A study of the solution of (8.84) is useful in more general problems just as the study of Airy's DE proved to be useful in the study of general turning point problems. See Exercise 8.8.

As usual, we first consider the reduced DE

$$xY' - Y = 0 , \ -1 < x < 1,$$

(8.85)

which has the solution

$$Y = Cx .$$

(8.86)

Examining (8.84) for boundary layers, we note that there can be no rapidly decaying solution near the left- or right-hand boundaries, since $2x$—the coefficient of y'—is, respectively, negative or positive in these regions. Yet if $\epsilon y''$ were small everywhere, we would have $y \approx Y$, which does not contain enough arbitrary constants to satisfy the BC's. As (8.85) has a singularity at $x = 0$, we would be justified in taking a different value for C in (8.86) for $x < 0$ than for $x > 0$. These values of C can be chosen so that both boundary conditions in (8.84) are met. Nevertheless, the resulting function is not a solution of (8.84) because its first derivative will be discontinuous at $x = 0$.

In the present example, the situation can be analyzed exactly by first noting that (8.86) is, in fact, an exact solution of (8.84). This reduces the determination of the complete solution of (8.84) to quadratures, via a reduction of order.

To simplify algebra, let

$$x = \epsilon\xi. \tag{8.87}$$

Then the DE in (8.84) reads

$$\frac{d^2y}{d\xi^2} + 2\xi\frac{dy}{d\xi} - 2y = 0, \quad -\epsilon^{-1} < \xi < \epsilon^{-1}. \tag{8.88}$$

Note that the scaling (8.87), while removing ϵ completely from the DE (8.88), has introduced ϵ into the solution domain. This is an illustration of the third point in the introduction to Chapter III.

Since $y = \xi$ is a solution of (8.88), we set

$$y = \xi u(\xi). \tag{8.89}$$

In terms of u, (8.88) reduces to

$$\xi u'' + 2(1 + \xi^2)u' = 0. \tag{8.90}$$

This is a first order linear DE in u'. Its solution is found easily to be

$$u' = c_1\xi^{-2}e^{-\xi^2}, \tag{8.91}$$

and so

$$u = c_2 - c_1 \int_\xi^\infty \frac{e^{-t^2}}{t^2}dt$$

$$= c_2 - c_1(\xi^{-1}e^{-\xi^2} - 2\int_\xi^\infty e^{-t^2}dt) \tag{8.92}$$

$$\equiv c_2^* + c_1^*[\xi^{-1}e^{-\xi^2}\sqrt{\pi} + \text{erf}(\xi)],$$

where $\text{erf}(\xi)$ is the well-studied and tabulated *error function*. Therefore, by (8.89),

$$y = c_2^*\xi + c_1^*[e^{-\xi^2}\sqrt{\pi} + \xi\text{erf}(\xi)]$$

$$= A_2x + A_1[(\epsilon\sqrt{\pi})e^{-x^2/\epsilon^2} + x\,\text{erf}(x/\epsilon)]. \tag{8.93}$$

Imposing the BC's in (8.84), we have

$$A = -A_2 - A_1 + \text{transcendentally small terms (TST) in } \epsilon$$
$$\text{(8.94)}$$
$$B = A_2 - A_1 + \text{TST in } \epsilon.$$

Thus,

$$y = x\left[\frac{B-A}{2} + \frac{B+A}{2}\text{erf}(x/\epsilon)\right] - \frac{B+A}{2\sqrt{\pi}}\epsilon e^{-x^2/\epsilon^2} + \text{TST.}$$
$$\text{(8.95)}$$

The sketch of (8.95) in Fig. 8.3 shows the existence of an interior or "corner" layer of width $O(\epsilon)$ centered at $x = 0$.

Exercise 8.7. Discuss the solution of

$$\epsilon^2 y'' + axy' - y = 0, \; -1 < x < 1, \; y(-1) = A, \; y(1) = B.$$

Exercise 8.8. Discuss the solution of

$$\epsilon^2 y'' + f(x)y - y = 0, \; -1 < x < 1, \; y(-1) = A, \; y(1) = B.$$

where $f(0) = 0$ and $xf(x) > 0, x \neq 0$.

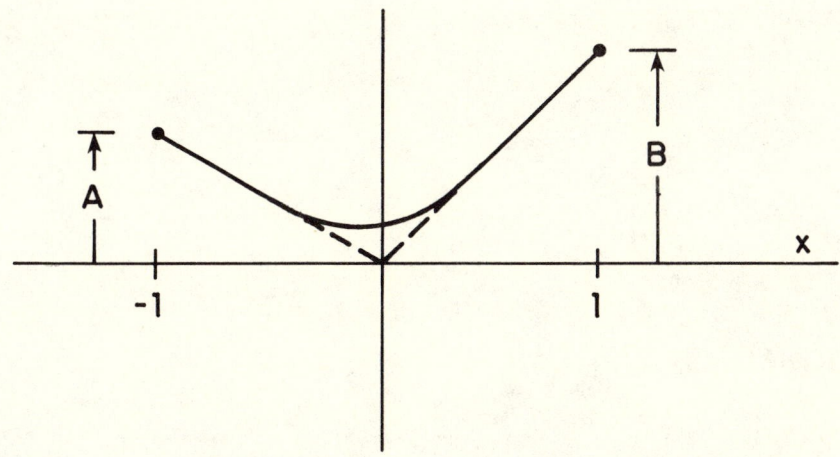

Fig. 8.3. The solution of (8.84) which exhibits a corner layer.

CHAPTER IX:
CABLES AND CELLS: ANCIENT AND MODERN PROBLEMS

Boundary layer methods in nonlinear problems were initiated by Prandtl near the turn of the century. Prandtl noted that when a fluid of low viscosity such as air or water flows about an obstacle, the ratio of viscous to inertial forces is small everywhere except in a narrow layer near the boundary of the obstacle. Using this observation, he was able to simplify considerably the analysis of the governing Navier-Stokes equations.

Problems of fluid flow are too elaborate for this book,[1] but Prandtl's ideas may be illustrated in other ways. To hint at the efficacy of boundary layer techniques for solving nonlinear problems, we have chosen two problems, one old, one new, both of practical importance we think, and both without exact, closed-form solutions.

The Shape of a Hanging Cable with Small Bending Stiffness. This problem springs from the much older *Catenary problem:* to find the shape of a hanging chain, *i.e.*, a cable with no bending stiffness. This might well be called THE PROBLEM of late 17th Century physics. Though now stated and solved in terms of the hyperbolic cosine in nearly every first-year calculus text, the problem challenged the greatest mathematicians of the time. For a fascinating account of these struggles see Truesdell, *The Rational Mechanics of Elastic or Flexible Bodies 1638–1788, L. Euleri Opera Omnia,* Series II, Volume II, Part 2, Zurich, Fussli.[2]

Consider a hanging cable with ends encased in rigid piers, as shown in Fig. 9.1. Let s denote distance along the cable from its low point and

---

[1]See the excellent, pioneering text by vanDyke, *Perturbation Methods in Fluid Mechanics,* annotated edition, the Parabolic Press, 1975.

[2]As far as we know, the analogous problem for a cable with bending stiffness—the full nonlinear version—was first set by one of us (JGS) on an examination in 1971 at the Technical University of Denmark. Shortly thereafter, a perturbation solution was published by A.M.A. van der Heijden, "On the Influence of the Bending Stiffness in Cable Analysis," *Proc. Kon. Ned. Ak. Wet. B76,* 1973, pp. 217–229.

$w(s)$ its weight/length. Further, let $H(s)$, $V(s)$, and $M(s)$ denote, respectively, the horizontal and vertical force and the moment exerted over the cross-section at s by the material to the right, as indicated in Fig. 9.1. Finally, with respect to a right-handed set of Cartesian axes Oxy, with O at the low point of the cable and the y-axis pointing vertically upward, let $x(s)$, $y(s)$ and $\phi(s)$ denote the Cartesian coordinates and tangent angle of a point P on the cable at a distance s. For the part of the cable between O and P, we have

Horizontal force equilibrium: $H(s) = H(0) \equiv H_0.$ (9.1)

Vertical force equilibrium: $V(s) = \int_0^s w(t)dt.$ (9.2)

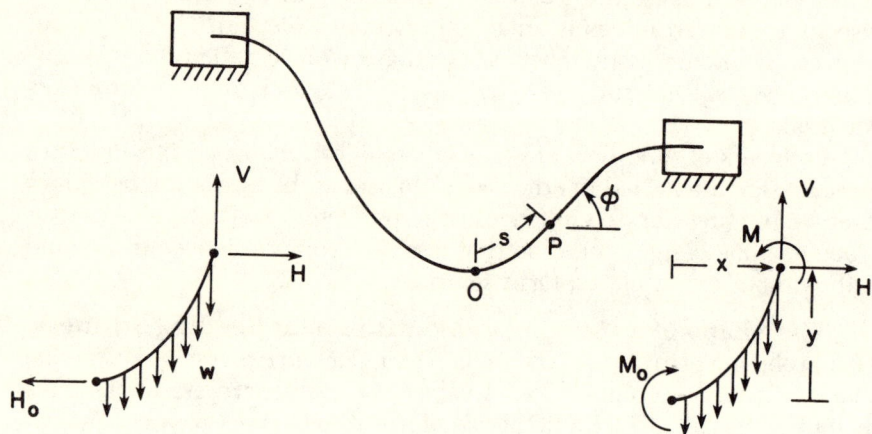

Fig. 9.1. The geometry of and the forces on a hanging cable with stiffness.

Moment equilibrium: $M(s) = M(0) - x(s)V(s) + y(s)H(s)$

$$+ \int_0^s x(t)w(t)dt. \qquad (9.3)$$

To round out our set of equations, we borrow from strength of materials the moment-curvature relation

$$M(s) = EI(s)\phi^{\cdot}(s). \qquad (9.4)$$

where $\phi^{\cdot} = d\phi/ds$. Here, as with the drill string of Chapter VII, E is Young's modulus and I is the moment of inertia of the cross-section. (Even though our problem is nonlinear, it may be shown that, so long as the cable is thin, a linear moment-curvature relation is quite accurate.)

We may reduce (9.1) to (9.4) to a second order DE for ϕ by differentiating (9.3), and, in the resulting equation, inserting (9.4) into the left side and (9.1) and (9.2) into the right side. Thus

$$(EI\ \phi^{\cdot})^{\cdot} = -x \cdot \int_0^s wdt + y \cdot H_0. \qquad (9.5)$$

But

$$x^{\cdot}(s) = \cos\phi(s)\ ,\ y^{\cdot}(s) = \sin\phi(s). \qquad (9.6)$$

Hence

$$(EI\ \phi^{\cdot})^{\cdot} + (\int_0^s wdt)\cos\phi - H_0\sin\phi = 0. \qquad (9.7)$$

We may restrict attention to the right half of the cable in which case (9.7) is to hold for $0 < s < L$, where L is the length of the right half of the cable. The BC's on ϕ are then

$$\phi(0) = \phi(L) = 0. \qquad (9.8)$$

Simplification and Nondimensionalization. Let us take E, I, and w to be constant and introduce the following dimensionless variable and parameters:

$$z = s/L\ ,\ k = wL/H_0\ ,\ \epsilon^2 = EI/L^2H_0. \qquad (9.9)$$

Then (9.7) and (9.8) take the form

$$\epsilon^2\phi'' + kz\cos\phi - \sin\phi = 0\ ,\ 0 < z < 1\ ,\ \phi(0) = \phi(1) = 0, \qquad (9.10)$$

where a prime denotes differentiation with respect to z. In terms of L and ϕ,

$$x = L\int_0^z \cos\phi(t)dt\ ,\ y = L\int_0^z \sin\phi(t)dt. \qquad (9.11)$$

The Catenary is the shape of a cable with no bending stiffness. The angle Φ of the Catenary is obtained by setting $\epsilon = 0$ in the DE in (9.10):

$$\Phi = \tan^{-1}(kz). \tag{9.12}$$

As is obvious from the physics, Φ fails to meet the BC at $z = 1$. Clearly our BVP has a singularity in the model.

Exercise 9.1. Determine x and y for the Catenary from (9.11) by replacing ϕ by Φ and verify that $y/l = \cosh(x/l) - 1$.

Boundary Layer Arguments. From experience, we know that if the cable is very thin ($\epsilon^2 \ll 1$), it must hang very nearly as a chain except near the ends. There ϕ must turn rapidly so that the cable can meet the pier horizontally. That is, ϕ must exhibit a boundary layer near $z = 1$. The dimensionless length 1 of the right half of the cable is one scale in our problem. The other is the width of the boundary layer, say δ. What is δ? We argue in rough terms as follows. From (9.12) we know that in the cable, just outside the boundary layer, $\phi \approx \tan^{-1} k = O(1)$. But as we move through the boundary layer, ϕ must decrease to zero to satisfy the BC. Thus ϕ undergoes a change of $O(1)$ over a distance δ, which suggests that, if ϕ changes smoothly, then ϕ' in the boundary layer is $O(\delta^{-1})$. But outside the boundary layer ϕ' as well as ϕ is $O(1)$, which means that ϕ' *itself* has undergone a change of $O(\delta^{-1})$ in moving a distance δ across the boundary layer. Thus, the derivative of the derivative of ϕ, ϕ'', is $O(\delta^{-2})$. But the DE for ϕ, (9.10), tells us that, *within* the boundary layer, $\epsilon^2 \phi''$ must be of the same size as $k \cos \phi - \sin \phi$, for if $\epsilon^2 \phi''$ dominated, then the DE would say that $\phi'' \approx 0$, *i.e.*, $\phi' \approx$ constant which, as we have just argued, is not possible, since ϕ must turn quickly within the boundary layer. On the other hand, if $k \cos \phi - \sin \phi$ were dominant in the boundary layer, then we would be back to the Catenary, (9.12), which does not satisfy the BC at $z = 1$. Thus, in the boundary layer, $\epsilon^2 \phi'' = O(\epsilon^2/\delta^2) = -k \cos \phi + \sin \phi = O(1)$, which implies that the width of the boundary layer is $O(\epsilon)$.

> Boundary layer (or, more generally, order of magnitude) arguments are potent, heuristic guides for sorting out the various scales in a problem. Their justification is usually *a posteriori*, in the apparent consistency and reasonableness of the approximations they suggest.

The two-scale method may now be applied, systematically, to reduce our singular perturbation problem to a regular one.

We introduce a fast (or boundary-layer) variable via

$$\zeta = \frac{1 - z}{\epsilon} \tag{9.13}$$

and assume that

$$\phi = \phi(\zeta, z, \epsilon). \tag{9.14}$$

By the chain rule and (9.13),

$$\phi' = -\epsilon^{-1}\phi_\zeta + \phi_z \tag{9.15}$$

$$\phi'' = \epsilon^{-2}\phi_{\zeta\zeta} - 2\epsilon^{-1}\phi_{\zeta z} + \phi_{zz}, \tag{9.16}$$

so that our BVP (9.10) takes the form

$$\phi_{\zeta\zeta} + kz\cos\phi - \sin\phi - 2\epsilon\phi_{\zeta z} + \epsilon^2\phi_{zz} = 0 \; , \; 0 < \zeta < \epsilon^{-1}, \; 0 < z < 1,$$
$$\phi(\epsilon^{-1}, 0, \epsilon) = \phi(0, 1, \epsilon) = 0. \tag{9.17}$$

We now assume that the introduction of the boundary-layer variable ζ has made ϕ regular in ϵ, *i.e.*,

$$\phi(\zeta, z, \epsilon) = \overset{0}{\phi}(\zeta, z) + \epsilon\overset{1}{\phi}(\zeta, z) + \cdots . \tag{9.18}$$

Substituting (9.18) into (9.17), using the Taylor expansions

$$\sin (\overset{0}{\phi} + \epsilon\overset{1}{\phi} + \ldots) = \sin\overset{0}{\phi} + \epsilon\overset{1}{\phi}\cos\overset{0}{\phi} + \cdots \tag{9.19}$$

$$\cos (\overset{0}{\phi} + \epsilon\overset{1}{\phi} + \ldots) = \cos\overset{0}{\phi} - \epsilon\overset{1}{\phi}\sin\overset{0}{\phi} + \cdots , \tag{9.20}$$

and equating to zero coefficients of like powers of ϵ, we obtain the sequence of BVP's

$$\overset{0}{\phi}_{\zeta\zeta} + kz\cos\overset{0}{\phi} - \sin\overset{0}{\phi} = 0 \; , \; 0 < \zeta < \epsilon^{-1}, \; 0 < z < 1$$
$$\overset{0}{\phi}(\epsilon^{-1}, 0) = \overset{0}{\phi}(0, 1) = 0. \tag{9.21$_0$}$$
$$\overset{1}{\phi} - (kz\sin\overset{0}{\phi} + \cos\overset{0}{\phi})\overset{1}{\phi} - 2\overset{0}{\phi}_{\zeta z} = 0 \; , \; 0 < \zeta < \epsilon^{-1}, \; 0 < z < 1$$
$$\overset{1}{\phi}(\epsilon^{-1}, 0) = \overset{1}{\phi}(0, 1) = 0, \tag{9.21$_1$}$$

etc.

Exercise 9.2. Work out (9.21$_2$).

A *First Integral* of the PDE in (9.21$_0$) is obtained by multiplying by $\overset{0}{\phi}_\zeta$. This enables us to write

$$\frac{1}{2}\left(\overset{0}{\phi}_\zeta^2\right)_\zeta + (kz\sin\overset{0}{\phi})_\zeta + (\cos\overset{0}{\phi})_\zeta = 0, \tag{9.22}$$

which implies that

$$\frac{1}{2}\overset{0}{\phi}_\zeta^2 + kz \sin \overset{0}{\phi} + \cos \overset{0}{\phi} = C(z), \; 0 < \zeta < \epsilon^{-1}, \, 0 < z < 1, \tag{9.23}$$

where $C(z)$ is an unknown function of integration.

The exact solution of (9.23) can be expressed in terms of elliptic integrals. However, a remarkable simplification occurs (no elliptic integrals!) if we first determine $C(z)$. Imagine fixing a point P on the centerline of the cable and then letting the stiffness approach zero. Clearly, the slope of the centerline at P will approach that of the Catenary. That is, if we *fix* z and let $\epsilon \to 0$, then, from (9.13), $\zeta \to \infty$ and

$$\overset{0}{\phi}(\zeta, z) \sim \Phi(z), \; \overset{0}{\phi}_\zeta(\zeta, z) \sim 0. \tag{9.24}$$

But (9.23) holds for all z in $(0,1)$ and all ζ in $(0, \epsilon^{-1})$. Thus, in particular, it must hold in the limit as $\zeta = \epsilon^{-1} \to \infty$, which implies that

$$C(z) = kz \sin \Phi(z) + \cos \Phi(z)$$

$$= \sec\Phi(z), \tag{9.25}$$

where the last line comes from (9.12).

To further reduce (9.23) let

$$\overset{0}{\phi}(\zeta, z) = \Phi(z) - \theta(\zeta, z), \; \theta > 0. \tag{9.26}$$

Then, with the aid of a little trigonometry, we find that

$$\theta_\zeta = -2\sin (\theta/2)\sec^{\frac{1}{2}}\Phi(z). \tag{9.27}$$

Exercise 9.3. Show the steps leading from (9.23) to (9.27).

To solve (9.27), separate variables and integrate:

$$\int \csc(\theta/2)d(\theta/2) = -\sec^{\frac{1}{2}}\Phi(z) \int d\zeta, \tag{9.28}$$

i.e.,

$$\ln \tan (\theta/4) = -\zeta\sec^{\frac{1}{2}}\Phi(z) + D(z), \tag{9.29}$$

where $D(z)$ is a function of integration.

The first BC in $(9.21)_0$ is satisfied automatically; the second, by (9.26), requires that

$$\theta(0,1) = \Phi(1). \tag{9.30}$$

Hence,

$$D(1) = \ln \tan\Phi(1). \tag{9.31}$$

Setting

$$D(z) = D(1) + E(z), \tag{9.32}$$

solving (9.29) for θ, and substituting the resulting expression into (9.26), we obtain

$$\overset{0}{\phi}(\zeta,z) = \Phi(z) - 4\tan^{-1}\{\tan[\tfrac{1}{4}\Phi(1)]\exp[-\zeta\sec^{\tfrac{1}{2}}\Phi(z) + E(z)]\}. \tag{9.33}$$

The function $E(z)$ cannot be determined without considering the BVP $(9.21)_1$ for $\overset{1}{\phi}$. However, since $E(1) = 0$, *we may ignore* $E(z)$ *in* (9.33) to within the error made *by approximating* ϕ *by* $\overset{0}{\phi}$. To within this same error, *we may also replace* $\Phi(z)$ *in* (9.33) *by* $\Phi(1)$.

Exercise 9.4. Justify these claims by showing that

$$\exp[-\zeta\sec^{\tfrac{1}{2}}\Phi(z) + E(z)] = [1 + O(\epsilon)]\exp[-\zeta\sec^{\tfrac{1}{2}}\Phi(1)].$$

With the aid of (9.12) and (9.13) and the approximations just mentioned, we may express $\overset{0}{\phi}$ as follows:

$$\overset{0}{\phi} \sim\tan^{-1}(kz)-4\tan^{-1}\{\tan(\tfrac{1}{4}\tan^{-1}k)\exp[\epsilon^{-1}(1-z)(1+k^2)^{1/4}]\}. \tag{9.34}$$

A graph of the right side of (9.34) for several values of ϵ and $k = 0.6$ is given in Fig. 9.5.

Exercise 9.5. A designer is interested in the direct and bending stresses in a cable, σ_D and σ_B. These are given by $\sigma_D = (H^2 + V^2)/A$ and $\sigma_B = (rE/L)d\phi/dz$, where A is the cross-sectional area and r is the radius of the cable. Taking $H_0 = 2000$ lbs., $w = 5$ lbs./ft., $r = 1$ in., $L = 200$ ft., and $E = 10^7$ lbs./in.^{2}, compute the maximum values of σ_D and σ_B.

A Problem from Cell Biology.[3] Our modern example is one which certainly cannot be solved analytically and would be troublesome to solve numerically. The problem is to determine the effect of a protein on the diffusion of calcium through the walls of the intestines. We assume that the protein forms a compound with the calcium and that both the calcium and the calcium-protein compound are transported by diffusion. The calcium concentration is constant on the inside of the membrane and is pumped out by a Michaelis-Menten pump at the outside membrane. The

---

[3]See Kretsinger, R. H., Mann, J. E., and Simmonds, J. G., "Evaluation of the Role of Intestinal Calcium Binding Protein in the Transcellular Diffusion of Calcium," *Proc. 5th Workshop on Vitamin D* (A. W. Norman, Ed.) 1982, pp. 233–248.

membrane is impervious to the protein: the protein cannot escape the cell. The following equations represent a mathematical formulation of the problem.

$$D_B \frac{d^2B}{dx^2} - k_{on} A{\cdot}B + k_{off} \overline{AB} = 0 \tag{9.35a}$$

$$D_A \frac{d^2A}{dx^2} - k_{on} A{\cdot}B + k_{off} \overline{AB} = 0 \tag{9.35b}$$

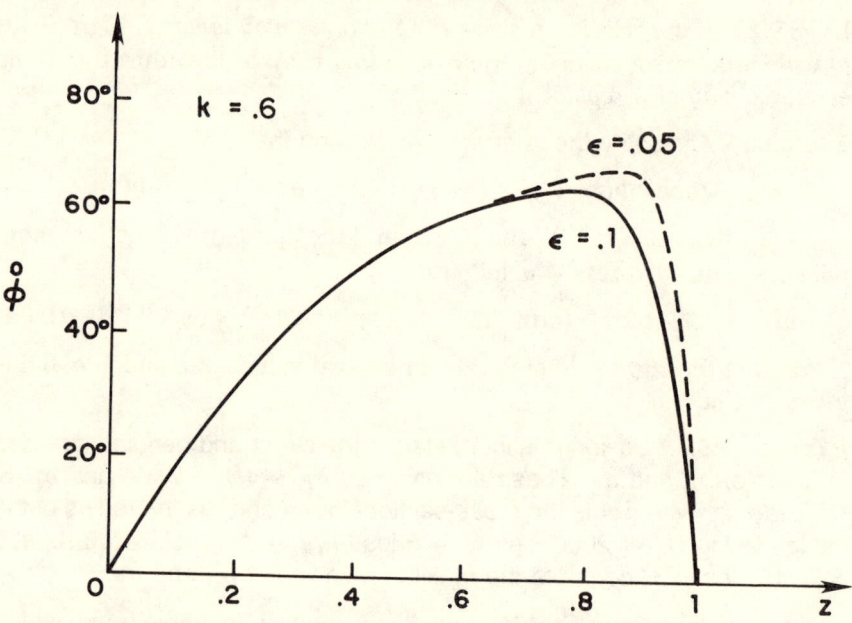

Fig. 9.2. Lowest approximation to the slope of the hanging cable.

$$D_{\overline{AB}} \frac{d^2\overline{AB}}{dx^2} + k_{on} A \cdot B - k_{off} \overline{AB} = 0 \qquad (9.35c)$$

$$B(0) = B_0, \quad -D_B \frac{dB}{dx}(L) = \frac{V_{max}B(L)}{B(L)+K_m} \qquad (9.36a,b)$$

$$\frac{dA}{dx} = \frac{d\overline{AB}}{dx} = 0 \text{ at } x = 0 \text{ and } x = L. \qquad (9.37a,b)$$

In these equations, A is the concentration of protein [mol/cm^3], B is the concentration of calcium, and $\overline{AB}$ is the concentration of the protein-calcium compound. The other parameters are identified in Table 9.1.

One final equation is required to specify the amount of protein in the membrane, namely

$$\text{Total Protein} = \frac{1}{L} \left(\int_0^L A\,dx + \int_0^L \overline{AB}\,dx \right)(\text{mol/cm}^3). \qquad (9.38)$$

These equations are simplified by introducing dimensionless variables and parameters as follows:

$$\xi = x/L, \ y = B/B_0, \ u = A/B_0, \ v = \overline{AB}/B_0 \qquad (9.39)$$

$$\lambda = V_{max}L/D_B B_0, \ \mu = K_m/B_0, \ k_1 = (B_L^2/D_A)k_{on},$$
$$k_2 = (L^2/D_A)k_{off} \qquad (9.40)$$

$$\sigma = D_B/D_A, \ s = \text{total protein}/B_0 \qquad (9.41)$$

Table 9.1. Cell parameters

Symbol	Nominal Value/Units	Remark
B_O	$3.16 \times 10^{-9} mol/cm^3$	calcium concentration at inside wall
D_A	$2.0 \times 10^{-6} cm^2/sec$	diffusion constant for protein
$D_{\overline{AB}}$	$2.0 \times 10^{-6} cm^2/sec$	diffusion constant for compound
D_B	$10^{-5} cm^2/sec$	diffusion constant for calcium
k_{on}	$10^{11} cm^3/mol\ sec$	rate of forming compound
k_{off}	$10^2 sec^{-1}$	rate of decomposition of compound
V_{max}	$5 \times 10^{-2} mol/cm^2 sec\cdot$	max pumping rate of Michaelis-Menten pump
L	$5 \times 10^{-3} cm$	thickness of membrane

In (9.39)–(9.41) we have taken $D_A = D_{\overline{AB}}$. Equations (9.35)–(9.37) now reduce to

$$\sigma y'' - k_1 uy + k_2 v = 0 \tag{9.42a}$$

$$u'' - k_1 uy + k_2 v = 0 \tag{9.42b}$$

$$v'' + k_1 uy - k_2 v = 0 \tag{9.42c}$$

$$y(0) = 1 , \ -y'(1) = \frac{\lambda y(1)}{y(1) + \mu} \tag{9.43a,b}$$

$$u'(0) = v'(0) = 0 , \ u'(1) = v'(1) = 0 \tag{9.44}$$

$$s = \int_0^1 u\,d\xi + \int_0^1 v\,d\xi . \tag{9.45}$$

Equations (9.42a,b,c) may be added to produce certain simpler equations. After some additions and integrations we arrive at

$$\epsilon y'' - [1 + A\gamma - Cy'(0)\xi]y - Cy^2 + D(s-\gamma) + y'(0)\xi = 0. \tag{9.46}$$

In this equation A, C and D are constants whose values are related to the original parameters; γ, on the other hand, is a constant of integration from (9.42a,b,c) which cannot be determined until the solution of (9.46) is known. Accompanying (9.46) are the conditions

$$y(0) = 1 , \ y'(0) = y'(1) , \ -y'(1) = \lambda y(1)/[(y(1) + \mu]. \tag{9.47a,b,c}$$

Equation (9.47b) is required to insure steady state conditions. From Table 9.1 we have

$$\epsilon = 8.0 \times 10^{-4}, \quad A = 0.632 , \ C = 3.16, \quad D = 0.200. \tag{9.48}$$

If we set $\epsilon = 0$ in (9.46), we get the reduced equation

$$-[1 + A\gamma - Cy'(0)\xi]Y - CY^2 + D(s-\gamma) + y'(0)\xi = 0. \tag{9.49}$$

This equation is simply a quadratic in Y containing the two unknowns γ and $y'(0)$; we want the positive solution. To proceed, let

$$Y(0) = L , \tag{9.50}$$

where L (which now stands for "left") is the solution of (9.49) at $\xi = 0$. Near $\xi = 0$ we set

$$y = Y + y_{BL} . \tag{9.51}$$

When (9.51) is substituted into (9.46), we find that

$$\epsilon y''_{BL} - [1 + A\gamma - Cy'(0)\xi + 2CY]y_{BL} - Cy_{BL}^2 + \epsilon Y'' = 0. \tag{9.52}$$

With $\tau = \xi\epsilon^{-1/2}$ as the boundary layer coordinate, we expand Y_{BL} in powers of $\epsilon^{1/2}$ to obtain, to lowest order,

$$\eta'' = (1 + A\gamma + 2CL)\eta + C\eta^2$$
$$\equiv P^2Q^2\eta + \frac{3}{2}Q^2\eta^2, \tag{9.53}$$

where $y_{BL} = \eta + \epsilon^{1/2}\eta_1 + \cdots$. The solution of (9.53) which goes to zero as τ goes to ∞ is

$$\eta = P^2\mathrm{csch}^2(\frac{1}{2} PQ\epsilon^{-1/2}\xi + \alpha), \tag{9.54}$$

where we choose α so that

$$1 - L = P^2\mathrm{csch}^2\alpha. \tag{9.55}$$

Equation (9.55) insures that $y(0) = 1$. It now follows that to lowest order [*i.e.*, to within a relative error of $O(\epsilon)$],

$$y'(0) \approx \frac{d\eta(0)}{d\xi} \tag{9.56}$$

$$= \frac{P^3Q}{\epsilon^{1/2}}\mathrm{csch}^2\alpha\,\cosh\alpha = \frac{Q(1-L)(1-L+P^2)^{1/2}}{\epsilon^{1/2}}.$$

On the right end of the interval, we introduce the boundary layer variable $\tau = (1 - \xi)\epsilon^{-1/2}$ and again expand y_{BL} in powers of $\epsilon^{1/2}$ to obtain, to lowest order.

$$\zeta'' = [1 + A_1\gamma - Cy'(0) + 2CR]\zeta - C\zeta^2$$
$$\equiv \overline{P}Q^2\zeta - \frac{3}{2}Q^2\zeta^2. \tag{9.57}$$

where $y_{BL} = -\zeta + \epsilon^{1/2}\zeta_1 + \cdots$ and $Y(1) = R$. We put the minus sign in the expansion of y_{BL} to insure that ζ is positive. The solution which decays when ξ decreases is

$$\zeta = \overline{P}^2\mathrm{sech}^2[\frac{1}{2}\overline{P}Q\epsilon^{-1/2}(1-\xi) + \beta]. \tag{9.58}$$

The value of β is obtained from

$$y'(0) = y'(1) = -\left.\frac{d\zeta}{d\xi}\right|_{\zeta=1} = \frac{\overline{P}^3Q}{\epsilon^{1/2}}\mathrm{sech}^3\beta\sinh\beta. \tag{9.59}$$

Finally, the pump condition is satisfied if

$$-y'(0) = \eta'(0) = \lambda[R - \zeta(1)]/[R - \zeta(1) + \mu]. \tag{9.60}$$

The unknowns, R, L, γ enter (9.55), (9.56), (9.59), and (9.60) in some complicated, algebraic way. The easiest way to find the unknowns is inversely. Proceed as follows. Choose a value for L ($0 < L < 1$). Use (9.49) with $\xi = 0$ to express γ in terms of this L. Use (9.56) to find $y'(0)$. Then β can be determined from (9.59). Finally, (9.60) is satisfied by choosing λ. In this procedure we must find R and check that $R = \zeta(1)$ is positive; if not then we need to choose a larger value for L. The final matched solution is

$$y = Y_s + \eta - \zeta + O(\epsilon^{1/2}). \tag{9.61}$$

With the aid of this solution, we may plot curves like Figs. 9.3 and 9.4.

The above problem, though complicated, illustrates the power of perturbation theory with matched expansions to solve nonlinear boundary value problems. The two-scale method of Chapter V could also have been used to obtain an approximate solution to the calcium diffusion problem and we urge the reader to do so.

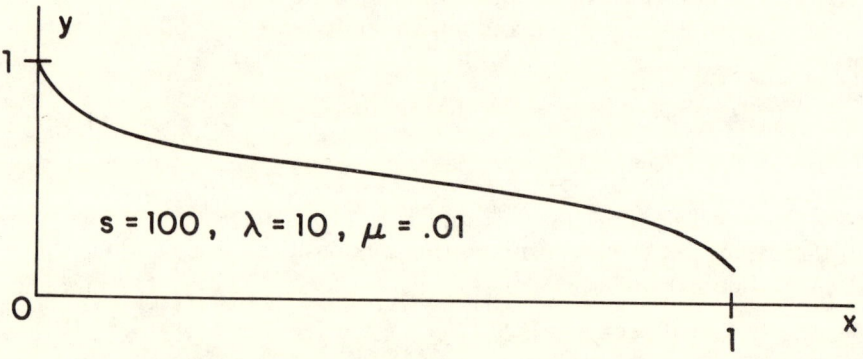

Fig. 9.3. Dimensionless Concentration of calcium vs. distance across the cell.

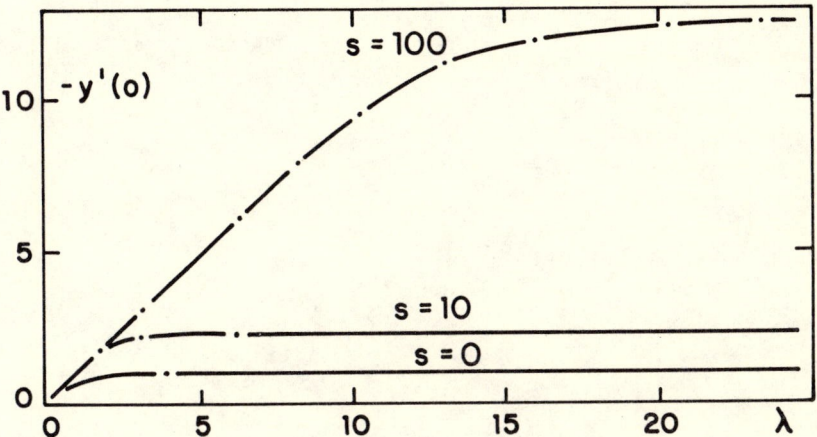

Fig. 9.4 Dimensionless flux of calcium vs. pump rate λ for various values of the dimensionless cell protein s.

BIBLIOGRAPHY

Here is a short list of books and articles, including some that have been mentioned in the text, that may be useful for reference or further study.

- Carrier, G., "Perturbation Methods," *Handbook of Applied Mathematics* (C. E. Pearson, Ed.), van Nostrand Reinhold, 2nd Ed., 1983, Chapter 14.

- Cole, J. D., *Perturbation Methods in Applied Mathematics*, Blaisdell, 1968.

- Erdélyi, A., *Asymptotic Expansions*, Dover, 1956.

- Kevorkian, J., and Cole, J. D., *Perturbation Methods in Applied Mathematics*, Springer-Verlag, 1981.

- Lakin, W. D., and Sanchez, D. A., *Topics in Ordinary Differential Equations*, Dover, 1982.

- Lin, C. C., and Segel, L. A., *Mathematics Applied to Deterministic Problems in the Natural Sciences*, Macmillan Publishing Co., 1974, Part B.

- Nayfeh, A. H., *Perturbation Methods*, Wiley, 1973.

- Nayfeh, A. H., *Introduction to Perturbation Techniques*, Wiley, 1981.

- Olver, F. W. J., *Asymptotics and Special Functions*, Academic Press, 1974.

- O'Malley, R. E., Jr., *Introduction to Singular Perturbations*, Academic Press, 1974.

- van Dyke, M., *Perturbation Methods in Fluid Mechanics*, annotated edition, Parabolic Press, 1975.

- Wasow, W., *Asymptotic Expansions for Ordinary Differential Equations*, Wiley Interscience, 1965.

APPENDIX A:
ROOTS OF $T_\epsilon(z)$ AND $T_0(z)$

The lemma on page 34 means: given any root z_k of $T_0(z)$ of multiplicity γ_k ($\gamma_1 + \gamma_2 + \cdots = n$), and given any constant $\rho > 0$ no greater than half the distance from z_k to the nearest distinct root of $T_0(z)$ (call it z_m), there exists a value of ϵ such that there are precisely γ_k roots of $T_\epsilon(z)$ in a disk of radius ρ centered at z_k. See Fig. A.1. A student who

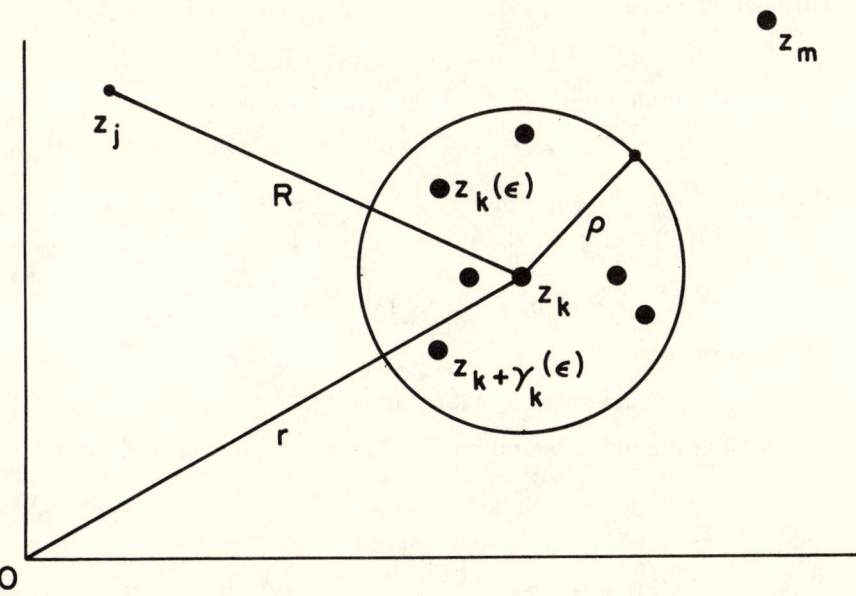

Fig. A.1. Some roots of $T_\epsilon(z)$ and $T_0(z)$.

has taken a first course in complex variable theory should have no trouble with the following proof.

Given a function $f(z)$ that is continuous and non-vanishing on, and analytic within, a simple, rectifiable, closed curve Γ, the number of zeros of f within Γ is, by Rouché's Theorem,

$$N(f) = \frac{1}{2\pi i} \int_\Gamma \frac{f'(z)}{f(z)} \, dz, \tag{A.1}$$

where $f'(z) = df/dz$. The idea is to show that if we take Γ to be a circle of radius ρ centered at z_k, then

$$|N(T_\epsilon) - N(T_0)| = \frac{1}{2\pi} \left| \int_{|z-z_k|=\rho} \frac{T_0(z)E'_\epsilon(z) - T'_0(z)E_\epsilon(z)}{T_0(z)[T_0(z) + E_\epsilon(z)]} dz \right| \tag{A.2}$$

will be less than 1 if ϵ is sufficiently small.

We do this by obtaining a lower bound on the magnitude of the denominator in (A.2) that does not vanish with ϵ and an upper bound on the magnitude of the numerator that does. The key is to note that the hypothesis $\lim E_\epsilon(z) = 0$ as $\epsilon \to 0$ implies two things: First, if ϵ is sufficiently small,

$$|E_\epsilon(z)| \leq (1/2)\rho^n \, , z \in \Gamma \, , \tag{A.3}$$

since z is bounded on Γ. Second, as $E_\epsilon(z)$ is of the form

$$E_\epsilon(z) = C_m(\epsilon)z^m + \cdots + C_1(\epsilon)z + C_0(\epsilon), \tag{A.4}$$

then, with $C(\epsilon) = \max |C_j(\epsilon)|, j = 1,2, \cdots m$ and $r = z_k$,

$$|E_\epsilon(z)| \leq |C_m(\epsilon)||r + \rho|^m + \cdots + |C_0(\epsilon)|$$

$$\leq mC(|r + \rho|^m + 1) \, , z \in \Gamma . \tag{A.5}$$

By the same reasoning,

$$|E'_\epsilon(z)| \leq m(m - 1)C(\epsilon)(|r + \rho|^{m-1} + 1) \, , z \in \Gamma . \tag{A.6}$$

To obtain a lower bound on $T_0(z)$ note from its factored form

$$|T_0(z)| = |z - z_1||z - z_2| \cdots |z - z_n| \tag{A.7}$$

that

$$|T_0(z)| \geq \rho^n \, , z \in \Gamma . \tag{A.8}$$

To obtain an upper bound on $T_0(z)$ and $T'_0(z)$, let $R = \max \{|z_j - z_k|, 1\}, j = 1,2,\ldots, n$. Then (A.7) implies that

$$|T_0(z)| \le (R + \rho)^n \,, z \in \Gamma, \tag{A.9}$$

Furthermore, as

$$T'_0(z) = (z - z_2)\cdots(z - z_n) + (z - z_1)(z - z_3)\cdots(z - z_n)$$
$$+ \cdots + (z - z_1)\cdots(z - z_{n-1}), \tag{A.10}$$

and as there are n terms each containing $n - 1$ factors,

$$|T'_0(z)| \le n(R + \rho)^{n-1} \,, z \in \Gamma. \tag{A.11}$$

Thus in (A.2)

$$|T_0(z)||T_0(z) + E_\epsilon(z)| \ge |T_0(z)|||T_0(z)| - |E_\epsilon(z)|| \ge (1/2)\rho^{2n} \tag{A.12}$$

and

$$|T_0(z)E'_\epsilon(z) - E_\epsilon(z)T'_0(z)|$$
$$\le |T_0(z)||E'_\epsilon(z)| + |E_\epsilon(z)||T'_0(z)| \tag{A.13}$$
$$\le C(\epsilon)[m\,(m - 1)(|r + \rho|^{m-1}+1)(R + \rho)^n$$
$$+ mn(|r + \rho|^m + 1)\,(R + \rho)^{n-1}].$$

Thus

$$|N(T_\epsilon) - N(D)| \le 2\rho^{1-2n}C(\epsilon)m(R + \rho)^{n-1}[(m-1)$$
$$\times (|r + \rho|^{m-1}+1)(R+\rho) + n(|r + \rho|^m+1)]. \tag{A.14}$$

But $\lim E_\epsilon(z) = 0$ as $\epsilon \to 0$ implies that $\lim C(\epsilon) = 0$. Thus, given any $\rho > 0$, the right side of (A.14) is less than 1 if ϵ if sufficiently small. Q.E.D.

APPENDIX B: PROOF THAT $R_{N+1} = O(\beta^{N+1})$

To prove that (4.35) is implied by the IVP (4.32), we first replace v in (4.32) by the right side of (4.34). Then, noting that λ has the form (4.33) and that $z_0, \cdots, z_N$ satisfy $(4.36)_0$ to $(4.36)_N$, we obtain an IVP for R_{N+1} of the form

$$\lambda^2 R''_{N+1} + R_{N+1} = \beta f(z_0, \cdots, z_N, R_{N+1}), \quad R_{N+1}(0) = R'_{N+1}(0) = 0. \tag{B.1}$$

By the method of variation of parameters, we convert (B.1) into the non-linear integral equation

$$R_{N+1}(T,\beta) = \beta \lambda^{-1} \int_0^T \sin \lambda^{-1}(T - \tau) f(z_0, \cdots, R_{N+1}(\tau,\beta)) d\tau. \tag{B.2}$$

The potential energy associated with (4.32) is $\frac{1}{2}v^2[1 - (1/12)\beta v^2]$. Hence (4.32) has periodic solutions if $\beta < 6$. By the choice of λ this period is 2π. But $z_0, \cdots, z_N$ are, by construction, 2π-periodic functions. Hence, by (4.34), R_{N+1} must be 2π-periodic. Thus, in (B.2), we may, without loss of generality, assume that $0 \leq T < 2\pi$.

The general method of proof may be inferred from that for $N = 0$, which we choose to minimize algebraic complexity. Thus, with $\lambda^2 = 1 + \beta\mu(\beta)$, $\mu(\beta) = O(1)$ and $z_0 = \cos T$, (B.2) reduces to

$$R_1(T,\beta) = \beta \lambda^{-1} \int_0^T \sin \lambda^{-1}(T - \tau)\{\mu(\beta)\cos \tau + [\cos \tau + R_1(\tau,\beta)]^3\} d\tau. \tag{B.3}$$

Taking absolute values of both sides of (B.3) and noting that all trigonometric functions are bounded in absolute value by 1, we infer the inequality

$$R \equiv |R_1(T,\beta)| \leq \beta\lambda^{-1} \int_0^T [|\mu| + (1 + R)^3] d\tau$$

$$\leq k\beta\lambda^{-1} \int_0^T (1 + R)^3 d\tau, \tag{B.4}$$

where $k = 1 + |\mu|$. Now (B.4) implies that $R(T) \leq S(T)$, $0 \leq T < 2\pi$, where

$$S' = k\,\beta\lambda^{-1}(1 + S)^3, \quad S(0) = 0. \tag{B.5}$$

The IVP (B.5) may be solved readily, in closed-form, by separation of variables. This yields.

$$R(T) \leq -1 + (1 - 2k\,\beta\lambda^{-1}T)^{-1/2}. \tag{B.6}$$

But k, λ, and T are $O(1)$. Thus, by (B.6) and periodicity, $R(T) \equiv |R_1(T,\beta)| = O(\beta)$, for *all* T. Q.E.D.

INDEX